ON MY MIND

Frontispiece
I am who I remember to be, the one I recognize each
day in the mirror, whose clothes are on the chair.
Memory is the seat of personal identity.
(*See* Reflections 1-24, 3-12, 5-13, 5-17, 5-23.)

ON MY MIND

MIND

*A New Vision
of Consciousness*

Steve Perrin

Lulu.com

O

Earthling Press
Bar Harbor

BY STEVE PERRIN

On My Mind (2013)

Consciousness: The Book (2011)

Acadia's Trails and Terrain (2002)

Acadia's Native Flowers, Fruits, and Wildlife (2001)

The Shore Path, Bar Harbor, Maine (2000)

Acadia: The Soul of a National Park (1998)

Myth to Mythology: Experience as a Resonant Synthesis of Meaning and Being (Dissertation, 1982)

Lulu.com ◯ Earthling Press, Bar Harbor, Maine 04609
On My Mind: A New Vision of Consciousness
Copyright © 2013 by Steve Perrin (Stephen G. Perrin)
All rights reserved. Published 2013
Printed in the United States of America
ISBN 978-0-9651058-2-8

Front Cover: Dreamself/dayself born *of* woman, born *to* the stars. 1) Francisco Goya, *El sueño de la razón produce monstruos*, (The dream of reason produces monsters), *Los Caprichos* (1799). 2) Anonymous wood engraving printed by Nicholas Camille Flammarion. Both courtesy of Wikimedia Commons.

Back Cover: Figure 2-5 reproduced in color. Tripartite model of the author's mind. The situated self mediates between perception and action. Diagram by the author. Accompanying text is excerpted from Reflection 7-15, Introspective reality, p. 175.

Every error in this book is my own. S.P.

In memory of

Laura A. Gale Perrin
August 25, 1864—September 17, 1896

*Despite the rain of Saturday, the
funeral of Mrs. Rev. J. N. Perrin, Jr.,
was largely attended. . . . The sad
and tender services of the hour were
closed with the baptism, beside the
open casket, of the son born two days
before, to whom was given the name
Porter Gale. At the age of a little more
than thirty-three years, the body of
Mrs. Perrin was laid to rest in our
village cemetery, her short life a long
one, having so nobly 'answered life's
great end.'*

CONTENTS (by Chapter & Reflection)

CONTENTS

CONTENTS

CONTENTS

CONTENTS

PREFACE

This book began a long time ago with an exploding star in a galaxy we now call the Milky Way. Its fuel expended, that star imploded into its own interior when gravity overcame its fading radiation. The collapse created a blazing forge that spewed atoms outward into space. Swirling as a cloud of gas and dust, those atoms circled a nameless star, coalescing five billion years ago into the third planet out from our sun. It didn't take those atoms long to form self-replicating molecules, to diversify and evolve into complex organisms, some becoming conscious of themselves, their activities, and their surroundings. Engaging those surroundings, one organism stood up, roamed around, and as a creature of the universe, developed individuality from a center of unique experience, coming much later to develop ideas, speech, writing, art, dance, books, and diverse societies.

Waking up in the morning, I discover myself to be one such animal, conscious of myself in this universe, of my surroundings, and of my engagement with those surroundings. I think to myself, this is hot stuff. Hot as in the inside of stars. Fiery. Too hot to get near, much less grasp. So I work for thirty years to understand how I, an assemblage of water, soil, and air in the presence of sunlight, can wake up in the morning with such a thought in my head. I don't understand how it is possible. Two years ago I brought out my first attempt at describing my mind in *Consciousness: The Book.* I learned a lot from that exercise and thought I could do better. So here is *On My Mind,* a second attempt to describe what it feels like for me to be conscious. My hope is to encourage others to make the same effort for themselves; then we can hold long conversations about the similarities and differences between us, and make a start at understanding how we can be conscious for the common good of this planet we all share together.

Steve Perrin
Bar Harbor, May, 2013

For Husserl the phenomenological method [suspending judgments and beliefs] demands that a philosopher place himself at a distance from all previously held theories and assumptions and become a nonpartici-pating observer of his conscious experiences of the world. This means that he cannot base his insights on traditional or well-established theories, whether philosophical or scientific, but on an immediate insight into the phenomena themselves. For a phenomenon precisely is one's immediate experience freed from all theoretical presuppositions and interpretations.

David Stewart and Algis Mickunas, *Exploring Phenomenology,* American Library Association, 1974, p. 36.

A phenomenology of consciousness [for the purpose of gaining insight into what conscious experience is like] is . . . relevant to a naturalistic investigation insofar as we need to have a good description of consciousness, to know what consciousness is, if, as psychologists or neuroscientists, we intend to explain how consciousness works or how it is generated.

Shaun Gallagher, Phenomenological Approaches to Consciousness, in Velmans and Schneider, eds., *The Blackwell Companion to Consciousness,* 2007, p. 686.

From the start, the dynamic generation of sympathetic rhythms is a bridge of "attunement" between the vitality of [infant and adult] minds.

D.N. Stern, *The Interpersonal World of the Infant,* 2000.

Only if we know what we are and why we are here can we decide how we should live together.

Adam Gopnik, Faces, Places, Spaces: The renaissance of geographic history. A Critic at Large, *The New Yorker,* Oct. 29 & Nov. 5, 2012, p. 121.

ON MY MIND

Chapter One

INTRODUCTION

1-1. Early comparisons. Since childhood, I have been fascinated by human differences. Differences between my mother and father, me and my elder and younger brothers. Between my teachers, my friends, my cousins, the neighbors I grew up with. For my entire life, every person I have ever met stands out as unlike any other. Comparison is a fundamental feature of my mind.

When I became an adult in my forties (I certainly was not one in college), I wanted to find out why two people undergoing the same experience would interpret it from such dissimilar perspectives. Driven by curiosity, I took a course at B.U. in personality theory. In my final paper I wrote that if, as the result of the day's experiences, a person did not wake up next day with a larger perspective on the world, then that person was not ready for a world that was sure to have changed overnight. My teacher gave me an A on the paper, a sign I read as encouragement for enrolling in grad school. In 1982 I received my Ed.D. after writing a dissertation exploring the comparative relationship between concrete and abstract dimensions of human experience.

Now it is thirty years later, and I've been reworking that dissertation in my mind the whole time through numerous bouts of introspection. Here are some of the reflections flowing from those decades of introspective experience.

1-2. Each mind is unique. Our personal genomes are unique, as are our immune and nervous systems, our childhood rearing and development, our educational experiences, our work histories, our situated perspectives on local and world events, our personal relationships, our autobiographies, our dreams, our streams of consciousness: seven billion minds, each unique.

1-3. *My* mind is unique. Any conclusions I draw about my mind through self-reflection may or may not apply to other minds. My authority for drawing conclusions at all stems from thirty years of probing what my mind has to tell me. The stories I base my findings on are told in narrative form depicting the drama of simultaneously living and observing a mental life. I am an observer—an experiencer—not a scientist pledged to particular methods of investigation and analysis.

1-4. Calibration. Our primal interactions shape our personal styles of engagement with persons-places-things by setting up the traffic patterns through the neural networks of our respective brains, thereby shaping the structure of those networks, and of our subsequent experience. We are products of our formative experiences in utero and early childhood. Subtly or shockingly, our families and communities teach us how to be human. We are calibrated by our families and communities to live a certain kind of life without realizing it. Like birds, as we are hatched to a particular nest, so do we sing.

1-5. Adaptability. Each of us is an experiment in adaptability to the unpredictable conditions we will meet later in life such as wars, recessions, new technologies, tsunamis, sickness, accidents, and countless other imponderables. Finding ways of adapting to the circumstances we are born to does not mean we are ready for the conditions we will grow into later on. It will take deliberate effort to refocus our engagements if they differ from those we were unwittingly trained to expect.

1-6. Introspection I. The most immediate access to our minds is through personal reflection on our own streaming consciousness, a job we must do for ourselves. Others can sometimes help us, particularly if they are empathic, but at the risk of confounding their thoughts and feelings with our own. Reflection implies an extensive period of close study or consideration, so introspection is no slap-dash enterprise of a moment or two (although insights can sometimes be gained in an extremely short time). It takes years of practice to familiarize ourselves with the full workings of our minds. No course of study is more beneficial, I now believe, than mapping our personal autobiographies onto our subjective worlds of experience.

The big hurdle to introspection is getting over the meta-phorical assumptions we are likely to make about what kind of mind we will find once we get there. Forget the computer idea, the clockwork, the screening room, the tabula rasa, the logical hierarchy, the jumble, the whimsy, the playground. Don't overlay a template of images on what you expect you will find. Be still and observe whatever comes. Stop, look, and listen. Pay attention from a noncommittal point of view to that which shows up. Then just follow along as best you can, second by minute by hour and day. There you will have it, your conscious mind at work and at play. And yourself taking note of yourself. There's always you the observer, and your mind as observed. Introspection engages both parties in some kind of mutual per-formance. By keeping track of both sides of the engagement, you will discover aspects of yourself as subject and object at the same time. And eventually see they are twin aspects of one unified life. Every river needs its banks and channel to make it a river. Slowly, slowly over the years, clarity will come. Don't push the river; let it flow as it will while you attend as you will. There you have it, introspection for a lifetime of close encounters between your coursing, complementary selves.

1-7. Responsibility. Because all of us are uniquely positioned to monitor the mind of only one person on Earth, that ready access predetermines—if the job of introspection is to be done—who has both the motive and opportunity to undertake the work in that case. The flow of awareness through any given mind is apt to vary so throughout the day, likely without notice, that only one person has not only opportunity but primary responsibility for watching over the sensory impressions, situated thoughts, behaviors, and engagements that emerge.

1-8. Elements of Consciousness. I have learned through thirty years of personal introspection that the seemingly unitary flow of my conscious mind comprises many elements (or dimensions) that shape any particular episode of consciousness according to a unique blend all its own. In this book I group those constituent elements under four headings: 1) perceptions, 2) felt situations, 3) actions, and 4) the resulting flow of engagement they maintain between my mind and my world (*see* Reflection 1-13 *below*).

The elements of consciousness I treat under *perception* include: expectancy, arousal, attention, sensory impressions, level of sensory discernment, and recognition (classification or interpretation). Those under *felt situations* include: comparison, understanding, creative imagination, feeling and emotion, biological values, acquired values, the life force, felt situations themselves, a sense of self as inner witness, speech, and memory (motor, working, autobiographical and conceptual). Those under *actions* include: judgments, decisions, goals, relationships, projects, skills, and demonstrable actions—the culmination of all that has gone before. Throughout, I emphasize a coherent sense of flow or engagement that unifies perceptions, felt situations, and actions into a serial progression characterizing my streaming mental activity.

Other investigators such as neuroscientists, therapists, psychologists, sociologists, and philosophers, analyze the components and shadings of the mind in different terms specific to their disciplines. The terms I use apply to what I find meaningfully descriptive of the ongoing narrative produced by my intuitive practice of self-reflection. This method gives me a clear picture of but a minute sample of the collective human mind.

1-9. Levels of Consciousness I. I discover my mind working on different levels of awareness, starting from a *reflex* level that happens so fast I am not aware of it; to a level of *mimicry,* on to a *routine* level obtained through extensive preparation and practice; an *habitual* level I achieve by repeatedly performing the same action, a *biased or prejudiced* level representing attitudes I neither question nor examine; a *dogmatic or ideological* level that, again, is not subject to personal monitoring; and finally the level of *full, deliberative consciousness* I achieve when facing problems that I do not know the answer to but feel a need to address.

These levels often shade one into another so subtly that I find it convenient to visualize my mind as occurring on four distinct levels: 1) reflex or imitative level; 2) routine or habitual level resulting from practice or repetition, leading to acting without thinking; 3) biased, dogmatic, or ideological level; and 4) the level of full consciousness.

The speed of these several levels of mental operation slows noticeably from the reflex level to full consciousness according to the mental effort it takes to perform them.

Complex behaviors can become automatic through repeated performance or rehearsal, converting demanding routines to reliable habits. This is how we perfect penmanship, for instance, or learn to play musical pieces on our instrument of choice without having to think consciously about what we are doing.

1-10. Vigilance. My mind is seldom quiet. On the go from one quest to another, it is kinetic and ever-changing. My job, the job of my situated self, is to try to keep pace as it shifts from one thought to the next. Thoughts often last for a mere fraction of a second, so if I want to capture them, I have to make a deliberate effort to recall the gist as best I can while it is fresh. I treat every thought as a fleeting glimpse of my interior state, so do my best to grasp what it has to tell me.

1-11. Independence. When young, I was taught to be well-mannered in deferring to others, that is, not to mind my own mind. Parents, teachers, brothers, friends—all wanted my attention. It was considered selfish and antisocial to think my own thoughts in their presence. Early on, I realized that to be independent and free, I needed to distinguish and keep a respectful distance between my thoughts and those situated in other streams of experience. On many occasions, that decision has made it possible for me to focus deliberately on the task of building bridges between my mind and, for example, those of family members, friends, colleagues, and students.

1-12. I am a phenomenologist. I think for myself, perceive for myself, build a world for myself. Not philosophically or conceptually, but experientially and existentially. The sensory impressions I entertain in my mind may stem from patterns of energy in the world, but the meaning and significance I assign to those impressions are my doing, based largely on comparison with what I can recall of my personal sample of lived experience. As a result, I have never been certain that the world *is* one way or another because I have no direct access to things and events as they might be in themselves. I have only such phenomena or impressions of those things and events according to one man's limited sensibility, sensory apparatus, and memory. I am fully aware that others are likely to form impressions that differ from mine. My senses do not relay "information" from the exterior so much as they create impressions tailored to my (unconscious) self-interests as best they can, using what they have to work with in the time allowed during my engagement with such matters. In truth, I often account for the world by making it up as a work of creative imagination, largely on the basis of prior experience as poorly represented in a fallible memory. Too, I strive to hold at a distance the influence of the culture I was born to in order to appreciate its calibration of my expectancies for, and understanding of, the mysterious outside world.

1-13. The big picture. Of the twenty-four elements or dimensions I distinguish in my own mind (*see* Reflection 1-8 *above*), I find it particularly useful when talking with others to sort those elements into four basic groupings, which I repeat here: 1) *sensory impressions,* 2) *felt situations,* 3) *actions,* and 4) *the flow of engagement* between those three and the world of matter and energy on the far side of my senses. (*See* Figure 2-1 *below.*)

1-14. Sensory impressions. I refer to the sensory patterns that appear in my mind (through sight, hearing, touch, taste, smell, and other senses) as *impressions* because they do not represent literal renditions of patterns in the world so much as such patterns as edited and emphasized by my sensory apparatus for my personal benefit. Generalizing, I would claim that we learn to see (hear, touch, etc.) what is important to us in terms of our personal life experience. I once set my hair on fire by reading too close to a candle flame; I have regarded the sight of candles in a cautionary light ever since. We say the sky is blue, but that is a function of pigments in our eyes translating diffuse radiant energy into a sense of blueness as distinct from redness or greenness. We say the stars overhead circle through the night sky, but we are the ones who circle with Earth's rotation, not the stars en masse, leaving our concept of the universe (which means *one turning*) as an unsupported or erroneous concept in our minds. If we own, say, a poodle, we are sensitized to the sight of other poodles, and are apt to look upon spaniels or huskies with less caring and discernment.

1-15. Felt situations. Our minds (here I am generalizing from personal experience) are situated in a particular moment of experience, each moment having a lineage of precursory moments, and leading to a legacy meted out as a series of subsequent moments. Our consciousness is located in that flow of moments, always having the fixed reference point of our *selves* as ever-vigilant observers. Each of us is the dreamer (thinker, perceiver, recaller) as our attention shifts from one moment to the next. As the officiating observer, we provide ever the same background to the scene passing before us in the foreground. Our awareness at each moment of experience includes both ourselves and the focus of our attention *as situated within the flow of our lives.* Where were you at 9:03 a.m. on the morning of September 11, 2001? I'll bet you can recall your situation at that time because strong emotions (fear, hurt, horror) made it memorable.

1-16. Speech I. Our talk gives voice to the situations we review or put ourselves in when we speak. Those situations provide the deep structure from which our words flow in response to the tensions inherent within them. The *structure of those situations* is the message (what it feels like to be me), not (as McLuhan claimed) the medium in which it is conveyed. In writing these words, I put myself in the situation of having to share the perspective I take in observing my own thoughts, for that is *where I am coming from.* I ask you to feel what it is like being me within that situation as you are able to recreate it in your own mind.

Trust, curiosity, good will, skepticism, self-knowledge—I have no control over such factors as may influence how you take in these words. The best I can do is make the call as I have witnessed it over thirty years of self-reflection, and lay it out for you to compare with your own introspective experience. If that comparison sparks your consciousness into grappling with where words come from in your own mind, that's probably as much as I can reasonably expect to accomplish. It is beyond me to try to convert you to my way of thinking or self-witnessing.

1-17. Actions. We act not so much *in the world* but (as we speak) *out of the situation we are in* at any given moment. Our behavior is situated in our minds more than in the world of matter and energy we construct around our sensory impressions. Once, never bothering to look against the flow of traffic on a one-way street, I stepped off a sidewalk into the road—and was felled by a bicyclist coming the wrong way, landing me in the path of a truck picking up speed after making a turn, which I thought I could beat, but as it turned out, the truck driver had to jam on the brakes to avoid running over my prostrate form in the road. The situation in which I acted was more complex than I realized, which almost cost me my life. I found I would be likely to live longer if I looked both ways before crossing even a one-way street to make a fuller assessment of my situation.

1-18. Time and space I. Both sensory impressions and physical actions change our awareness, the former as compared to a cultural standard of time and duration, the latter compared to a cultural standard of distance and direction, so our loops of engagement course through space-time at every cycle they make, updating and locating us again and again. Which sounds all very abstract and philosophical, but experientially is as concrete as bacon and eggs in that through sensory impressions and our behavior we actually participate in the experiential medium of space-time. Our culture provides the standards, but our awareness and our actions provide the specific perceptual and behavioral changes we hold against those standards, so placing ourselves within a framework of time and space. With the result that our experiential units of time are synchronized in different minds, as is our sense of space around and within us. Movers and shakers—those who take the initiative, whether in voyages of discovery, sports, battle, or for that matter sex—exist in a framework of space (*here, here,* and *here*), while those shaken or moved-upon exist in a framework of time (*now, now,* and *now*). When I write, time doesn't exist for me; when I listen to music, space doesn't exist. As I put it in *Consciousness: The Book:*

> A good portion of my concern with sensory phenomena is about detecting change in my environment (trucks moving in my direction, for instance), as well as changing phenomena resulting from movements I make while walking or turning my head. These two sorts of changes have different signatures in consciousness because in observing the one I hold myself still (as in a seat at school or in a theater) and in the other I must [unconsciously] subtract effects due to my own movement from the resulting sensory patterns in order to navigate through the world and be clear about what I see. I think of these two kinds of phenomenal changes as *it*-changes and *self*-changes. An it-change, for instance, would be the apparent round of the sun through the sky from east to west [and back again] in about twenty-four hours. A self-change would be my walking through woods while avoiding branches and roots. Calibrating it-changes in seconds, minutes, [etc.], they become measurable against a standard of *time.* Calibrating self-changes in inches, feet, yards, and miles, they become measurable against a standard of *space.* To go further, our concept of time is based on it-changes set as standard measures of duration; our concept of space is based on self-changes set as standard measures of our range of motions (it is no accident that our English system of measurement is based on dimensions of monarchal body parts). The import being that our two

most fundamental systems of measurement are based
on calibrated (that is, finely categorized) phenomenal
changes in human consciousness. Time and space do
not exist in the universe apart from human consciousness. They are our mental schemes projected outward
for the purpose of calibrating the universe in terms
meaningful to humans, and humans alone (Lulu.com &
Earthling Press, 2011, pages 35-36).

1-19. Loops of engagement I. The actions we perform come back
to us via proprioception (which gives us a sense of our posture
and bodily position in space) and the other sensory impressions
we form in the aftermath of even our most subtle behaviors
(such as directing our gaze). That feedback completes what I call
the loop of engagement, connecting us to the material and energetic
world via both our behavior and, on the receptive side, the
impressions we form as a result of our directed action and
attention. That give-and-take maintains a perpetual looping
exchange of behaviors and sensory impressions at the heart of
our conscious awareness, keeping us informed regarding the
effectiveness of our actions, setting the stage for what we are to
do in the next moment. Such engagement is far more consequen-
tial than mere hand-eye coordination or a streaming succession
of working memories. After thirty years, I now realize that this
dynamic—as set off by the comparison of my hopes and expectations against the results I actually achieve—is the engine that
drives my personal consciousness. (*See* Figure 2-5 *below.*)

1-20. Organ systems I. All of our organ systems engage their
surroundings much as the nervous system does through a
comparable loop of engagement, though largely unconsciously.
That way our bodies adjust themselves to our physical surroundings, achieving homeostasis within the range of conditions
we can tolerate. Our pulmonary and cardiovascular systems
cooperate to exchange oxygen and carbon dioxide, reproductive
systems assure engagement between egg-and-sperm-bearing
partners, integumentary systems assure support and renewal of

our skin, immune systems do battle with foreign invaders, and digestive systems meet both the nutritional and waste-removal needs of our bodies—all on an ongoing basis, thereby enabling our lives and activities. The loop of nervous engagement is no exception to the complex exchange that even single cells must transact through their outer membranes in engaging with their surroundings to sustain a viable range of inner conditions supportive of life.

1-21. Nervous system. Our individual nerve cells all partake in engagements required for their individual life support. But, too, the nervous system as a whole processes energy received from other systems, as well as from natural, cultural, and social worlds outside our bodies, determining the situations such energy represent to us, and deriving more-or-less suitable courses of action appropriate to each situation as it develops. The mind does not deal with the outside world directly as it might exist in itself. I have never met electrons, atoms, or molecules such as are alleged to compose the material world. Instead, I can only construe, construct, or infer (on the basis of training or hard-won experience) such a world as derived from the impressions I form as prompted by the signals I receive, and the meanings I elaborate on the basis of those impressions in light of my prior learning and experience. Those meanings give structure to my life world—the situation—that I find most compelling and relevant at the moment, and to which I am challenged to make an appropriate and effective response in my further engagements.

1-22. Broken engagements. In my blog I told the story of a three-year-old girl I know who

> got several presents at Christmas: toy dinosaur, plastic sow and piglet, miniature pocketbook with handle. She quickly saw the former two would fit snugly in the latter, so used the pocketbook as a shelter for her new treasures, which she proudly carried about the house. Until the handle came off, the pocketbook fell to the

floor, and [she] broke into tears, her momentary joy come to a sad end.

Ever the observer, I saw that episode as an example of a loop of engagement kindled in the mind of a child who, heartbroken when the pocketbook fell apart, cried in frustration and disappointment. As Steve Jobs cried when designers at Apple failed to live up to his demanding expectations for new product lines. (www.onmymynd.wordpress.com, Reflection 325: Stormy Weather, Sept. 28, 2012.)

Nothing demonstrates how much we depend on our abilities to engage than how we feel when a loop in progress comes to an abrupt end. Stressed, in a word. Thwarted. Frustrated. Disappointed. Crushed. Defeated. Helpless. Powerless. Hurt. Desperate. Rejected. Angry. Vengeful. Often expressed in tears, four-letter words, grunts, groans, curses, oaths, and other expressions of sudden displeasure. We're on a roll, and when we lose it, we're not happy. If we can't get started again, we get depressed.

On the other hand, when we succeed "beyond our wildest dreams," we're affirmed, gratified, elated, in seventh heaven, victorious, happy, powerful, successful, on top of the world, feeling secure, not threatened. When seriously engaged, there's no middle ground; we're up or we're down, we win or we lose. This points to our engagement with sports and games, conflicts, court cases, college classes, jobs, social relationships, business contracts, political allegiances, religious groups, and every other aspect of life we hold as important. Our everyday engagements include eating, shopping, traveling, sex, working, playing, planning, watching, listening—in even broader terms, participating in a fulfilling life. When our engagements get stalled, life takes a dive for the bottom.

The middle position on the three-way switch in our heads that turns consciousness off and on is where our unconscious minds take over and we can put our engagements on cruise control or automatic pilot so we don't have to pay deliberate attention to what we are doing. Consciousness activates when we are surprised by the degree to which awareness either *exceeds or falls short of our expectations,* so becomes salient; either way, we are engaged. Our unconscious minds don't call attention to themselves—they simply do their jobs and present us with the results as a *fait accompli.* As a child insists, "I can do it all by myself," and under routine circumstances, our minds do just that.

1-23. Consciousness begins with comparison. Much as depth perception is built from the disparity between images cast on the retinas of our left and right eyes as juxtaposed and compared in adjacent columns of our visual cortex, so, I propose, might consciousness arise from disparities between our situated expectations and what actually transpires in current awareness. Such expectations are abstracted from memories of similar occasions in the past, and serve as the norm or personal standard to which incoming sensory impressions are compared. If the two signals are similar, we are able to make a ready response. But if they are disparate, indicating that our impressions exceed or fall short of our expectations, then we become aroused, pay attention, and make a conscious effort to account for the difference so we can take appropriate action. In this respect, I visualize the brain as a vigilant comparator looking for the then in the now, and when not finding it, taking pains to update memory through conscious scrutiny. In that sense, consciousness is *a heightened state of mental activity* during which sections of the neural network can be modified and made relevant to novel, surprising, or unfamiliar situations.

My conjecture is that the loop of engagement begins at a site in the prefrontal lobe of the brain that feeds forward to centers of motor activity, courses (by means beyond our control) through the world outside our bodies, and ends at a different neural site as a terminus for highly processed sensory signals located in the parietal lobe. In our brains, those starting and ending points are linked by a variety of sites in which aspects of outgoing and incoming signals can be compared, resulting in a judgment that expectations have been 1) met, 2) exceeded, or 3) not attained, the relative discrepancy serving to initiate the next round of engagement, starting in the prefrontal lobe.

Scientists—in measuring their data, controlling their parameters, and keeping track of their variables—are expert comparators (measurement is a kind of comparison to a standard), so train themselves to be consciously watchful over their research and professional reputations. Politicians and soldiers are ever watchful over the claims and doings of those who oppose them. Parents worry about their children's safety, so are ever on the lookout for situations that contain potential threats. Churchmen compare human behaviors to precedents set forth in holy writ as standards to strive toward. To be vigilant is to monitor loops of engagement very closely, so to be ready to adjust a course of action at a moment's notice.

I view measurement, science, mathematics, speech, art, music, and aesthetics as being based on relative, mental comparisons between sensory elements as told (gauged) by their similarities or differences. The word "gauge" and the activity of gauging are among my favorite concepts because they so often describe the workings of my mind as I discover it through introspection. I regard my brain as serving as a gauge (comparator) of my world, allowing me to judge the similarities and differences between aspects of my experience. Whether calibrated in specific units or simply noting that something exists in greater or lesser degree than something else, my brain serves me more as a comparative gauge than what today we call a computer. I think more in terms of relative valences and relationships than computed differences. I observe symmetry directly with my senses instead of having to measure and compute it. Mathematical operations are unnecessary when tuning a piano or guitar, it is the sensitive ear that hears the beats when matching one note against another. Aesthetics gauges the intuitive relationships between sensed qualities such as line, texture, color, motion, direction, size, pitch, loudness, tone color, tang, scent, and so on. Such relations are immediately sensed through comparative experience; there is no need to compute them.

1-24. The situated self. The self is the superintendent of one particular mind. Its job description includes perceiving, remembering, acting, thinking, dreaming, comparing—whatever needs attending to. The self directs attention and recollection to bear on situation after situation, so enabling the mind to assimilate the now to its version of the then, or to accommodate that then to the now by modifying its neural connectedness to allow new ways of accounting for novel events. A felt situation incorporates the self as an observer (informed by memory, feelings, and values) *with* the scene as observed in the now. It allows us to watch ourselves riding the leading edge of our own lives as viewed from inside. The self, that is, is always situated in the stream of its conscious experience as set off against current sensory impressions, prior and potential actions, and memories from the past. This allows us to serve simultaneously as both subject and object of our own attentiveness, combining inner and outer worlds into a single world of situated awareness. I view that feat of synthesis as made possible by synchronous or conflictive electrochemical traffic through various regions of our brains, producing such patterns of comparative resonance and dissonance as elicit consciousness itself.

1-25. The brain is not the mind. I think it unlikely that the brain is aware of how the mind takes or interprets its neural activity any more than microbes in our gut are aware of their role in digesting our food. The brain's job is to provide an ample network of routes to support sufficient neuro-electrochemical traffic to maintain the worthiness of the mind (both conscious and unconscious) in meeting any and all situations upon which its survival may depend. As I picture it to myself, consciousness resides in the collective and interrelated flow of traffic throughout the brain, not solely in the material routing of that traffic from one brain area or cortical column to another. The mind as I see it results from comparison between different signals in the brain, not from particular signals in themselves.

1-26. Aesthetics I. Aesthetics is the comparative study of what we find attractive or repellant, how we describe such judgments, and what physical or emotional responses we make as appropriate. From my perspective, this puts aesthetics at the heart of consciousness as facing directly into the issue of what stimulates consciousness in the first place. Certainly we pay attention to sensory impressions and their relationships that strike us as either attractive or repulsive. We safely ignore what moves us neither way—that's just business as usual. What attracts or repels us, however, serves as a gateway to consciousness in arousing a state in which we pay deliberate attention to how we are affected, and to the challenge of making an appropriate response. In saying "there's no accounting for taste," or "beauty is in the eye of the beholder," we acknowledge that we are calibrated in different ways so that our conscious minds are apt to respond differentially to similar patterns of sensory stimulation. Therefore we find changing fads and predilections within ourselves, not agreed-upon laws or even principles of either aesthetics or of consciousness.

1-27. Learning. We learn by assimilating more-or-less familiar sensory patterns to old (established) neural networks, or by adjusting (altering, expanding) such networks to accommodate unusual patterns of sensory stimulation. Our understanding is either large enough to account for new episodes of experience, or must be wrapped around episodes we don't yet understand, growing larger in the process. If we don't exercise our minds in rigorous fashion very often to retain what we have learned, they may grow lax, and attrition of neural connections robs us of what we once thought we had mastered or understood, resulting in unlearning or forgetfulness.

Much of what I studied in school (based on what influential educators thought I should know) ended up as a sense of nostalgia (or gratitude) for what I once knew but have had few occasions to recall since then. I remember studying the imports and exports of South American countries, naval stores, and even differential equations—but absolutely nothing about how my mind worked or how I might use it effectively.

"Education" stems from the Latin word *educare*, meaning to draw or lead out. My schooling was more about laying concepts from other people's minds onto my native repertory, not developing the curiosity I brought with me when I walked into the classroom as a unique individual. I remember in second grade sitting in the back of the room, looking up, counting the number of perforations in the ceiling tiles (*see* Reflection 4-1 *below*). I also remember being shocked in seventh grade when, on the same day, we studied Mercator projections in two different classes, social studies and math, allowing for an overlap between what had been presented as disciplines in hermetic isolation. Such connections just were not planned, much less stressed, during my early schooling.

1-28. Fields of learning. Our traditional fields of endeavor and learning (such as physics, mathematics, history, agriculture, art, crime, technology, sociology, philosophy, homemaking, governance, religion, and so on) are truly *fields of cultural engagement* by which we learn how to direct our attention and acquire skills to attain the results we desire. We situate ourselves within one such field or another to become the kind of person we want to be (or others want us to become) by adopting the traditions, terminology, methods, skills, tools, accessories, and attitudes we think we need if we are to live up to our subjective aspirations. By situating ourselves within one such discipline, we ally ourselves with an identifiable segment of our culture, and so grow larger than our individual selves by acting in concert with a cohort of trainers, peers, colleagues, supporters, or gang members. We rise above our child selves to adopt a professional station in this life, donning the fitting uniform, framing the appropriate degrees and certificates on our walls, showing the appropriate scars, achieving the status we desire. All this requires retooling our neural networks to accord with the ideal self we aspire to grow into. That ideal bears some resemblance to the image we formed of particular adults we looked up to in childhood. So does the world achieve a vestige of kinetic stability (an oxymoron with bearable tensions) from one generation to the next, naïve possibilities becoming fulfilled as attainments through practice, dedicated study, and engagement. Thus do we experientially pull ourselves up and ahead by tugging on our own bootstraps.

1-29. Reality. I devoted Chapter 13 in *Consciousness: The Book* to a discussion of reality. Quoting from that work:

> Consciousness . . . depends on a working mind finding itself in a stimulating, energy-rich surrounding situation, both self and situation engaging in an ongoing exchange that can endure for one human lifetime. That exchange itself constitutes the reality of the two taken together as what we call consciousness. Reality resides neither in the person nor in her surroundings, but in the bioenergetic interaction between the two operating in tandem through the looping engagement they establish with each other. That, in essence, is the upshot of my introspective research. Reality comes down to our forming a secure relationship with our surrounding milieu, our niche in the universe, whether it is composed of significant others, work opportunities, energy sources, or situations in which we can make ourselves happen in the face of difficulty or opposition. Essentially ecological, reality is our term for the dynamic exchange of energy between inner and outer [worlds] that keeps us undead.
>
> And it is the disparity between those inner and outer states of energy (as revealed by our ability to compare them via our looping engagement) that gives rise to consciousness as an error signal waking us to our current reality. It is the bite of that error signal that alerts us to a discrepancy in our understanding of our situation, rousing us to consciousness so we can investigate the source of that signal which runs counter to our expectations (pages 219-220).

I do not subscribe to a concept of reality split between matter and mind as two different substances. Rather, I experience a cooperative and dynamic reality that acknowledges the necessary comparison between different neural signals representing the ionic and chemical flow of traffic within pathways in the brain. Arousal by an equally physical comparison between resonant and dissonant activity in different regions of the brain gives rise to a sense of the situated self and its mind. The mind is not contained in that resonance itself, but, I would say, is an emergent property of that resonance in relation to a contrasting area of dissonance that sets it off. If not a succinct explanation of how consciousness may arise, that is the story I tell myself.

1-30. Cultural consciousness I. In addition to mimicry and commanding authority, we have many ways for bringing separate minds into harmony with one another. Judges rule over proceedings in courts of law, Robert's Rules of Order govern countless meetings, consensus works for some, intimidation and displays of brute force for others. A good part of anthropology is devoted to cultural relativism and the mediation of differences. Political campaigns lead up to Election Day when voters favor the stance of one candidate over all others. We are in this world together, and devote a tremendous amount of time, energy, and resources to figuring out how we are to engage peacefully with one another, while having an impressive repertory of both peaceful and violent means at our disposal. Humor is an effective way to achieve accord between those who can laugh at themselves. Nothing stimulates engagement more than a smile, or inhibits it more than a furrowed brow.

1-31. Making ourselves happen. Because each of us is unique, our minds are our own. Our impressions, situated selves, actions, and engagements belong to us and no one else. There are many who would have us do *their* bidding, but we accept their offers at some personal risk—the potential sacrifice of the inherent, biological integrity of our actions, perceptions, and selves situated in our own minds and memories. If we do not stand for ourselves, who then do we stand for? We are made to engage by following our impressions with felt situations, our situations with meaningful actions. We are born to make ourselves happen in the world through engagements meaningful in relation to our goals and expectancies. That is what it means to be a person.

Our genome, our neural network, our rearing, education, work experience, and aspirations have meaning primarily because they affect the lives we actually lead. That is, they are consistent with who we are, so they are primarily meaningful in the context of our lives. Taken out of that setting, they may have meaning for someone else—whoever has taken over our bodily functioning and efforts. There is no shortage of bidders for our attention, skills, votes, beliefs, appetites, assets, allegiance, and bodily strength. Sometimes others will pay us good money for them, sometimes they will try to take them from us against our wishes. We are born to this world only once; if we find ourselves born a second time, it may suit someone else's needs and agenda more than our own.

1-32. Animal consciousness I. Do we see intelligence in the eyes of a wolf, dog, cat, ape, or monkey? That is, do those eyes—as direct extensions of their brains—reveal a capacity and willingness to engage us personally as fellow beings? Cats strike me as being a borderline case in that their eyes are not all that expressive, but they're good at catching birds, squirrels, and mice, and they like warm places to sleep, so we welcome them among us as our agents in performing such duties. Dogs as pets and companions have proven their eagerness to catch Frisbees, go on walks with us, bark at strangers, herd sheep, point at birds, sniff out drugs, and assist the blind in getting around, so we obviously engage with them and they with us by the terms of an assumed if unwritten contract. Caged primates sometimes reveal a hurt in their eyes at being forced to engage behind bars on *our* terms, not theirs. Unless, that is, they can convince themselves that they are the ones who are free, and it is their viewers who are held captive on the far side of those bars. Coyotes and wolves follow our movements with a wariness that betrays intelligence as if it were a matter of life or death—which it is.

The comparisons that spur consciousness require memory of prior experiences against which to place events in the now to see how they measure up. I am convinced that I see signs of such comparisons in the eyes of canines and primates, ever ready to approach, stand, or flee, depending on whether the comparison bodes good or ill from their situated points of view. Just because their repertory of behaviors may not be as rich as ours does not deny animals a form of consciousness similar to ours in being based on loops of engagement triggered by arousal, curiosity and comparison. Our sense of smell being much diminished by our rising up on our hind legs, we are unable to appreciate the richness of *being there* in our animal neighbors' experience.

1-33. Media of human exchange. Money generally passes as *the* medium of human exchange (as if there were no others), but kindness, civility, empathy, social equality, and language also qualify for that honor as effective playing-field levelers and stimulants of interpersonal engagement. Guile, fear, force, power, dominance, and submission are frequently used to skew human engagements, but only between unequal partners in cases where it is acceptable for one party to best another, as in sporting contests, warfare, and finance. Our situated selves decide whether or not such engagements are fair to the parties involved, leading to appointment of referees, umpires, regulators, judges, clerks, moderators, chairpersons, ombudsmen, and other engagement-enhancing officials.

I consider the elevated position of financial transactions and obligations in the economy-minded society I live in to be nothing less than a disorder of human engagement, which reduces all values to what they're worth in monetary terms. I see this emphasis as the result of engagements between people in responsible positions who put personal profit above good social order and public well-being. These are people likely to repeat their performance when they next get a chance. I regard them as sociopaths who profit from the lax ethical standards they have set for themselves.

Equally damaging disorders of engagement include autism, hyperactivity, dyslexia, amnesia, aphasia, allergies, post-traumatic stress disorder, attention deficit disorder, Alzheimer's disease, sensory impairment, physical impairment, drug-child-spousal abuse, cruelty, greed, depression, neglect, and chronic frustration.

1-34. Ownership. I view ownership as an outgrowth of the sense of territoriality by which we (and a great many other species) meet survival needs through skillful engagement with surroundings that can provide what we want. We feel we "own" what we frequently engage with, our situated selves regarding it as "mine." This applies to tools, accessories, pets, children, spouses, friends, abodes, properties—whatever serves us in our personal engagements. Which puts our "possessions" in a new light. It's not that we pay for them, but that they play a role in defining who we are and how we act. We "own" them in the sense they have become part of our routine. We "need" them in order to feel familiar to ourselves in conducting our engagements in ways we have grown accustomed to.

"Owning" property, for instance, puts us at odds with the obvious fact that we are products of the Earth and have no ownership rights over that which provides for us—any more than a robin can own the tree and surrounding territory it nests in for its "exclusive" use. Yet courts of law go to great lengths in defending our rights to parcels of land we pay money for, as if people owned the Earth and not vice versa. We agree to empower money to (symbolically) do that for us, an example of wishful thinking carried to an extreme degree. Clearly, we don't know our own minds well enough to appreciate how foolish we are in collectively setting a going rate on Earth's limited resources.

That confusion about ownership illustrates the kind of interpretive disparity that in 1978 sent my then forty-seven-year-old self off on a quest to get to the bottom of why people view similar sensory patterns so differently. Thirty-five years later, I now see that "being there"—wherever we are—entails engaging the Earth with deliberate mindfulness and care because our home planet is clearly our ultimate situation in this universe. If we don't place ourselves *there,* we are truly misguided, lost even to ourselves. I would say that the human experiment with consciousness is flawed in leaving judgment to the experimental subjects themselves—namely us. It turns out that many of us have never employed our creative imaginations to fully grasp the urgency of our worsening plight.

1-35. Being conscious together. Having come this far on my introspective journey, I see now that my situated self is at the balance point between two departments of my mind: 1) the one that *makes myself happen* in the world, and 2) the one in charge of *being there* in that same world. My awareness is situated, that is, between my own acting and being acted upon, between giving and receiving, doing and being done unto. The situated self that I am bridges two parts of my mind, one active, the other more observant, the three together making up the loop of engagement by which I deliberately connect and am connected to a world I can never know as it is in-and-of itself.

What I have been trying to do in this introduction is come to grips with the mental economy by which I exist in that world by acting and being acted upon. The *I* of myself and the *me* of myself are two parts of the whole person I am in looping succession as both subject and object of my own engagements with life. I am both driver and passenger, server and recipient of food, seller and buyer of goods, talker and listener, toucher and the one touched, lover and beloved, seer and one who is seen.

The challenge of living a life requires finding a balanced connection between the perceiving and acting parts of myself. I am neither strictly active nor passive, I can be both in alternation. I can stand on stage and sit in the audience, be eloquent and be attentive, teach and learn, talk and listen.

The situated self at the center of each mind is the one who joins the two halves in meaningful ways so they make sense in relation to each other. The situated self is the one who remembers to be him- or herself on any and all occasions. And, when the occasion demands, has the ability to expand the mind that he or she is to become a larger, more experienced, more understanding, more able person. But this transformation does not happen automatically. The situated self has to be in a position to encourage its core memory to change and grow larger, its self-image to become more accepting and expansive.

Each of us is a single cell adrift in a vast ocean. We have no choice but to drift where the current takes us. Having a metabolism that burns calories to provide the heat required for life, we each stay on the lookout for food, for companions and allies, for dangerous situations, for conditions in which we can thrive. For billions of years, across thousands of miles, our ancestors have been adrift on such currents—until they have brought us to this exact location where we find ourselves today, such as we have become at this hour. Across the billennia, we have changed in response to our altering situations, growing larger and more complex to meet the demands of our personal journeys. What defines and sustains us is our adaptability to the rigor of the journey we have made. If that rigor had killed off our ancestors when young, we would not be here today—and the same law applies to the relationship between ourselves and our far distant children of the future. If we succumb before sexual maturity, they will never be born, dying as possibilities before their time. Which is bound to happen sometime. The question is, on whose watch?

The question I have been addressing in these pages is: What am I that I discover myself living the life that I do? Based on personal self-examination, I have attempted to present various ways of considering my own mental life in response to that question. By way of example, I offer my personal loop of engagement as comprising my sensory impressions, situated self, and actions directed toward the world, the talented trio whose close harmonies have sung the life I have lived up till now.

Which leaves the question: How are we to live together amid the challenges we collectively face? How, indeed, are we to turn toward such a massed display of folly and absurdity? Only by addressing that display one-at-a-time on our own can we get past it to the next checkpoint on our path, as our children and grandchildren must do on their paths.

Here I offer what I have learned in my time as a prompt to remind those who follow me to do for themselves in their own time. Looking inward to know yourself is one way of facing into external difficulties so that you don't compound them and make them worse by behaving inappropriately to your true situation. What I have written here makes sense to me, situated as I am, not in the traditions of philosophy, psychology, or neuroscience, but in my own life and times. My legacy is the merest possibility of learning to know oneself, however it can be done, not a recipe describing how to do it step-by-step. There is no single way to acquire self-understanding. But I sincerely believe that if we don't try to develop it on our own, we may wait and wait, but never find it delivered to our doorstep within our lifetime. O

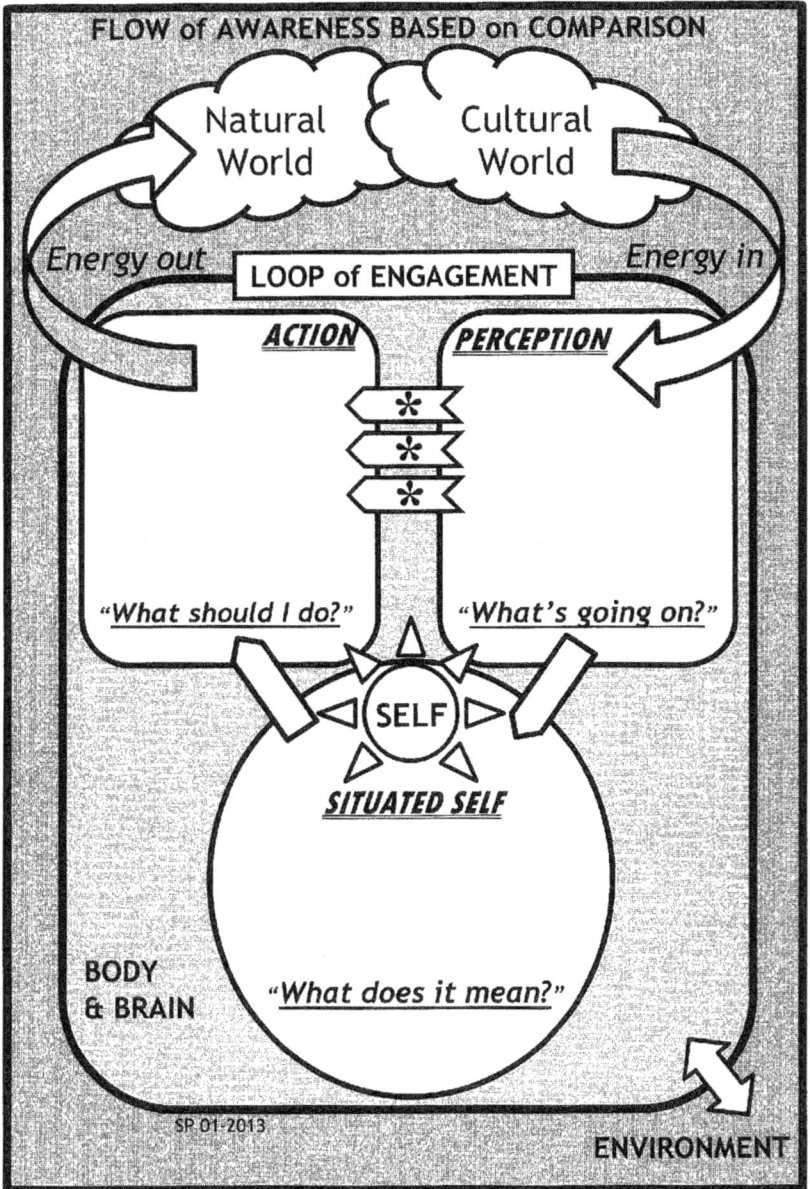

Figure 2-1. Three tasks, my mind's basic plan.

Chapter Two

DEPICTING A MIND

Figure 2.1. Three tasks. Having lived with my mind for eight decades, I find its labors sort naturally into three major tasks:

1. *Perception* converts patterns of ambient energy affecting my sensory receptors to impressions in my mind. Perception addresses the question, *What's going on?*
2. *My situated self* compares sensory patterns against patterns in memory, evaluating the import of each situation in relation to the query, *What does it mean?*
3. *My actions* deal with my thoughts and sensory impressions in as appropriate a manner as I can manage at the time, revealing to the world *What I feel moved to do.*

My mind springs from my brain, not the world as it is, from electrochemical traffic through my network of some hundred billion nerve cells, not from any outside world of daylight, fresh air, and presumed information. If I can't know the world as it is, at least my body and brain give birth to a mind that does the best it can to construe a world it can't know directly. Which is why I say I invent my world inside-out on the basis of whatever clues my mind can glean from the traffic through its brain.

What I do have is opinions, conjectures, and interpretations aplenty, based on the ambient energy impinging on my sensory apparatus as fitted to my understanding and personal experience, but not to "knowledge" residing in other people's minds. That is for them to know, and me to find out as best I can.

This figure suggests two different routes for my mind to travel from perception to action, one (diagonal arrows) by way of the self in the situation it consciously finds (or puts) itself in at the time, the others (horizontal arrows on several levels) passing unconsciously and directly to spontaneous action.

The large rectangle represents my mind's physical incarnation in its body and brain.

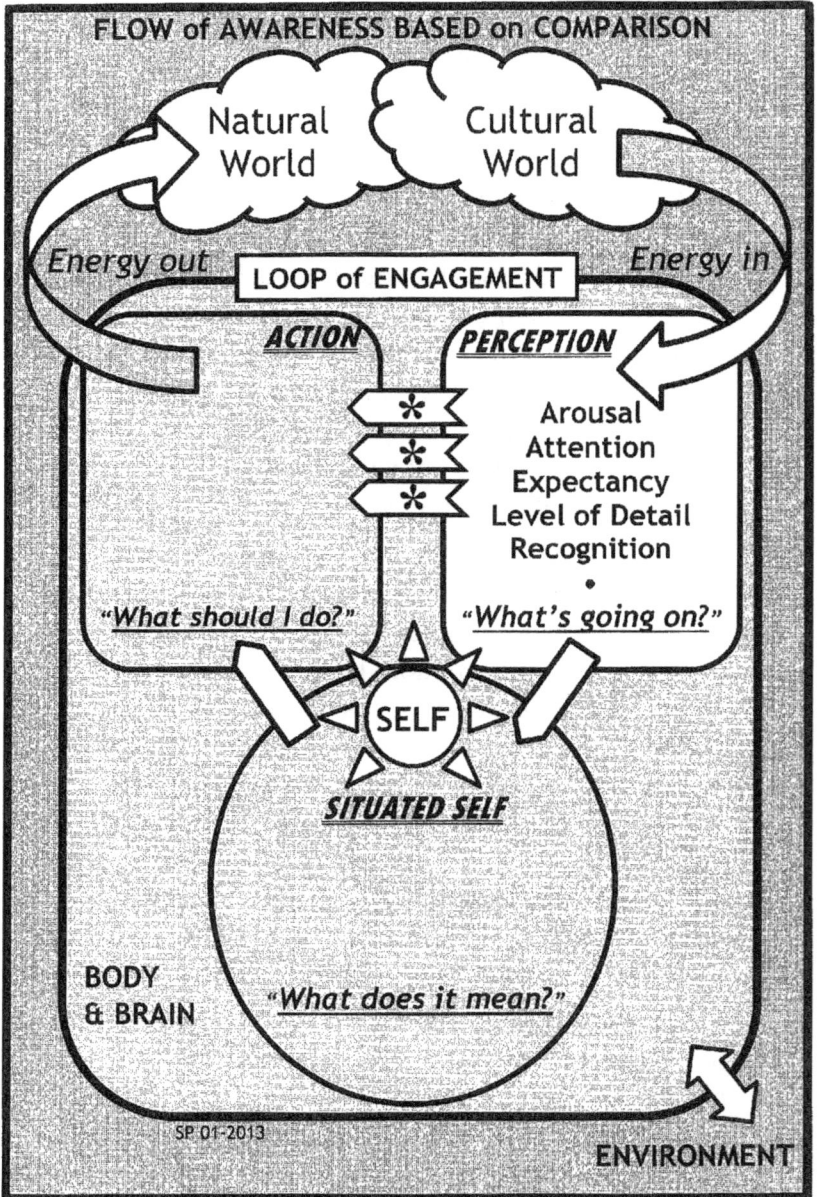

Figure 2-2. Perception alerts me to *what's going on.*

Figure 2-2. Perception. Perception entails a related cluster of mental dimensions which I may or may not be aware of:

1. *Arousal* from coma, sleep, or distraction resulting in a state of wakefulness in which attention is under willful or at least habitual control.
2. *Attention* to sensory stimulation, thoughts, memories, and other aspects of mental activity and experience.
3. *Expectancy* as governed by vestiges of prior experience in similar situations, or by wishful thinking and fantasies about what might conceivably happen under, for example, imagined, feared, or desired conditions.
4. *Level of detail,* acuity, discrimination, or discernment at which a given sensory impression is resolved.
5. *Recognition* of a sensory impression as belonging to a familiar and labeled conceptual category whose members are identified by a similar set of attributes, which I also refer to as *categorization* or *interpretation.*

Our sensory systems do a considerable amount of editing in shaping and presenting the impressions we become aware of. They heighten contrasts, link areas of similar attributes, overlook minor inconsistencies, and generally enhance perceptions along lines that conform to features we are familiar with and may be looking for. I have mistaken trash bags by the side of the road for dying crows, a cedar tree blowing in the wind for a man scraping paint, and snow on the Rocky Mountains for clouds blocking my view of those very mountains. I have learned not to take a rose as being a rose unless I can verify my judgment call; makers of artificial flowers stay in business because they are clever. Just fitting the right sensory pattern does not mean it's the real thing. Ask any maker of camouflaged clothing, padded shoulders, breast implants, or, again, a professional artist or actor.

Perception, as I see it, entails not only stimulation, but a reaching *for* stimulation, so that our impressions depend, in the end, both on what we're looking or listening for, and any stimuli within range.

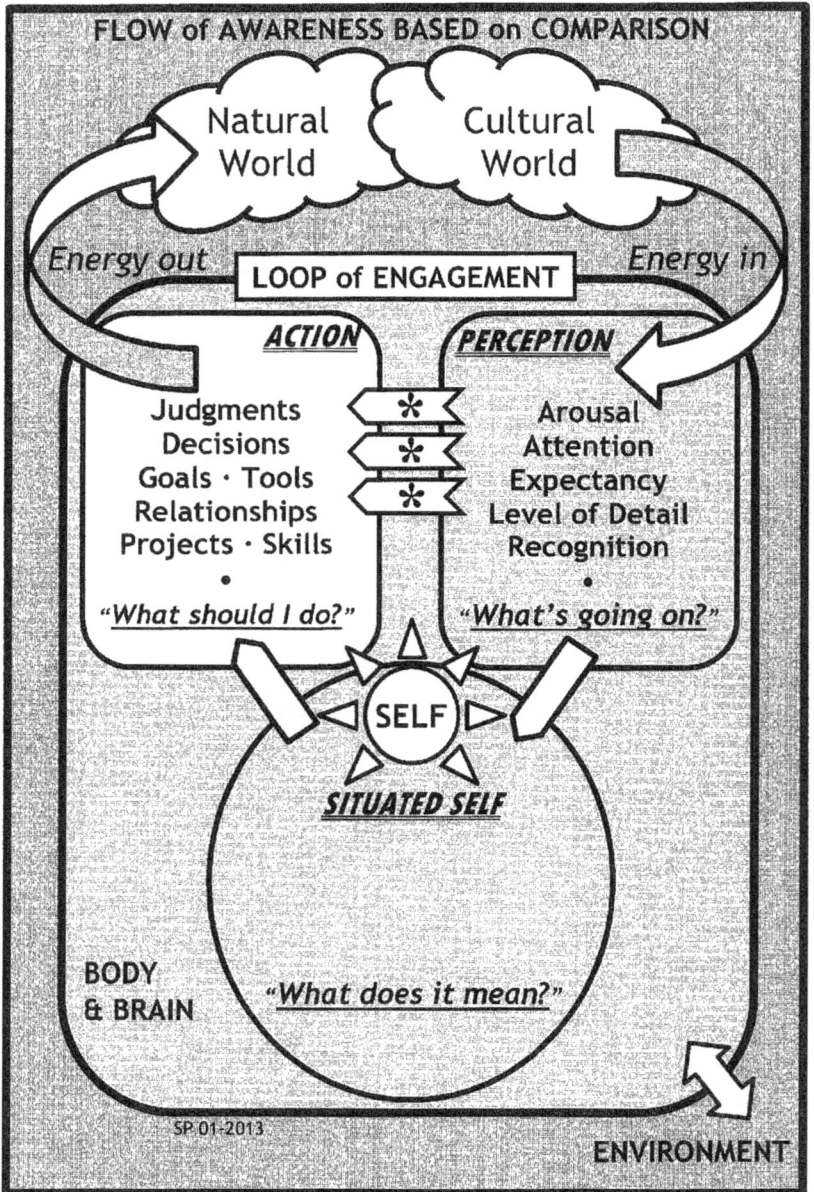

Figure 2-3. Action answers the query, *What should I do?*

Figure 2-3. Action. Since seeing is so often believing—as magicians, special effects departments, plastic surgeons, and package designers prove to us every day, we often act directly on the basis of our initial sensory impressions without giving conscious consideration to what we are doing. That's the danger those horizontal arrows are meant to point out, that reflexively acting on first impressions, bypassing full consciousness (represented by the oval), is often riskier than we assume. Action is best taken after a series of preparatory planning stages:

1. *Judgments* about the suitability of what we are about to get ourselves (and others) into.
2. *Decisions* regarding how we are going to go about what it is we want to do.
3. *Goals* specifically set to take us to the brink of action.
4. *Tools and accessories* chosen for their utility and effectiveness in producing the results we want to achieve.
5. *Relationships* we need to rely on in getting the job done.
6. *Projects* we will undertake to work our way through all the details we need to master if we are to pull it off.
7. *Skills* we need to do the job to our specifications.

It's not what we say we want to do but what we actually accomplish that determines the quality of our engagements. Résumés are sheets of worthless paper if they can't be followed up by concrete, predictable results. Every engagement is just one phase of the life we want to lead; follow-through is everything in getting to the phase beyond that. I remember watching movie comedian Joe E. Brown drive a caterpillar tractor over a wooden bridge, each plank falling into the water, leaving no way for him to get back. Each engagement implies the one after that. Life is an endless process, not a steady state. So shooting yourself in the foot may be funny to some, but not the marksman himself.

Projects are where our actions take shape day-by-day. Each has a rhythm of its own, which we adhere to in order to get the job done within budget and on time. The end point of consciousness is effective action in the world, particularly in the case of actions we've never taken before.

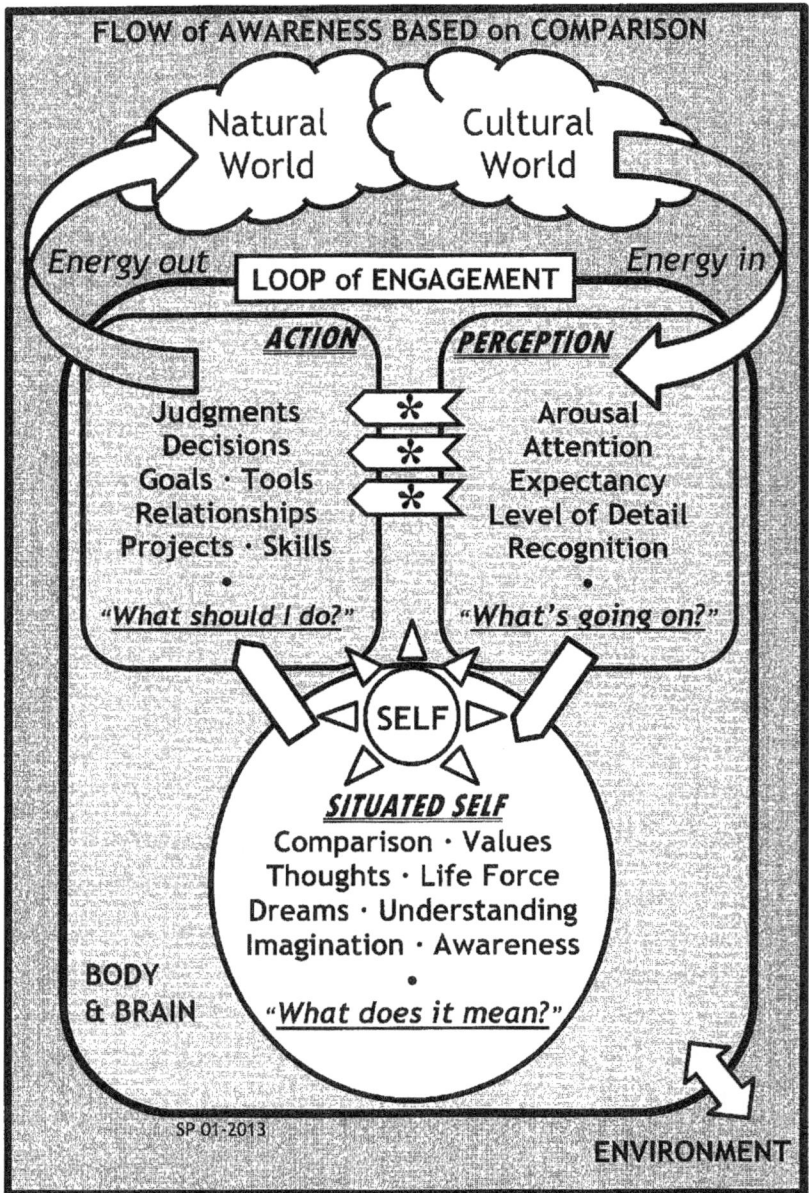

Figure 2-4. The situated self *gives meaning to experience.*

Figure 2-4. The Situated Self. Consciousness gives us an alternative to instinctively acting in the world as we have acted before. Consciousness frees us from merely mimicking others or acting on impulse, enabling us to transcend our accomplishments up until now. If our actions don't measure up to our hopes and expectations, we can still exercise our creative imaginations in making ourselves happen in exciting new ways:

1. *Comparison* is where consciousness begins, with a mismatch signal alerting us that our recent attainments either exceed or fall short of our expectations, enabling us to dig in and do better, or to raise our standards to a new height of accomplishment.
2. *Values* come in two sorts, biological or survival values, and those we acquire from the experience of living a life.
3. *Thoughts* are the quantum particles that pass through our minds, the basic units of consciousness. They can be sensory impressions, visualizations, memories, hopes, fears, desires, verbal phrases, dreams—the stuff of our particular and situated mental life.
4. *Dreams* are night thoughts by the dreamer who is the same self as the daytime actor, without being able to engage the world through action or perception.
5. *Understanding* is the field of personal learning in which we assemble all our conscious quantum particles according to their relationships as we see them fit together into a coherent and comprehensible whole.
6. *Imagination* goes beyond what we understand by playing with familiar events, so putting them together in unusual forms and new ways that excite us by transcending the world order we have grown accustomed to.
7. *Awareness* is what we not only direct our attention to, but also appreciate in a conscious manner. It entails being aware of thoughts and other mental phenomena, while being aware of ourselves being aware of our own situation or placement in our universe of consciousness.

The self is situated within the possibility of its mental activity.

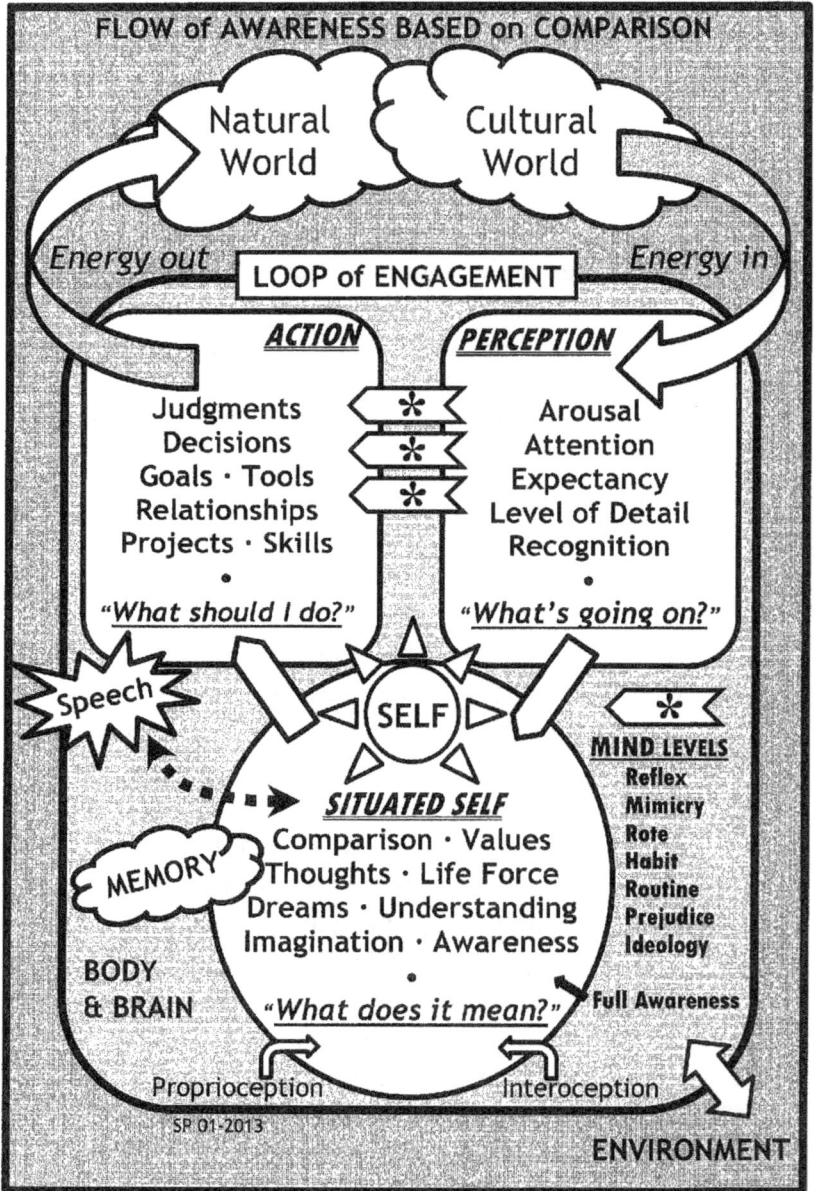

FLOW of AWARENESS BASED on COMPARISON

Natural World

Cultural World

Energy out   LOOP of ENGAGEMENT   *Energy in*

ACTION   *PERCEPTION*

Judgments
Decisions
Goals · Tools
Relationships
Projects · Skills
·
"What should I do?"

Arousal
Attention
Expectancy
Level of Detail
Recognition
·
"What's going on?"

Speech

SELF

SITUATED SELF
Comparison · Values
Thoughts · Life Force
Dreams · Understanding
Imagination · Awareness
·
"What does it mean?"

MEMORY

BODY & BRAIN

MIND LEVELS
Reflex
Mimicry
Rote
Habit
Routine
Prejudice
Ideology

Full Awareness

Proprioception   Interoception

SR 01-2013

ENVIRONMENT

Figure 2-5. Loops of engagement connect minds to worlds.

Figure 2-5. Loops of engagement. This is a full-dress portrait of my mind as I have grown familiar with it over thirty years of self-reflection. *Flow of Awareness Based on Comparison* refers to the disparity between expectancy and actual experience, the spark that ignites consciousness. I include the following dimensions:

1. *Speech,* as I feel it rising within me, originates in the deep structure of the synchronized, neural energies coursing through my brain at any particular moment. I have long wondered where words come from when they flow from my lips as I start to speak. Where am I, then,—where is the self of my mind—when that happens? Invariably, in one particular situation or another, which structures my concerns, so when I speak, my words address the tensions and relationships shaping that situation.

2. *Memory* flows from the structure of the network that conducts neural traffic through my brain, the electro-chemical resonances and dissonances within which reminders of specific episodes spring to mind.

3. *Mind levels* are here indicated by the three horizontal arrows linking perception directly to action, bypassing *Full Awareness,* which occurs within the oval, province of the self in its current situation. The levels I include are listed in the caption on the lower right side of the figure.

4. *Proprioception* refers to input detailing the physical position of head, torso, and limbs in relation one to another, providing a sense of positions in space necessary for synchronizing our bodily actions, and in reaching out within a framework of three-dimensional space.

5. *Interoception* refers to input from within our own bodies, in this case mine, conveying notice of aches and pains, pleasure, balance, feelings of contentment or disturbance, and subtle intimations felt in the "pit of my stomach."

6. *Loop of Engagement* refers to the streaming flow of mental traffic along the relatively easy (because unconscious) "high road" [horizontal arrows] or more difficult (fully conscious) "low road" [diagonal arrows] of awareness.

Chapter Three

PERCEPTION

3-1. Manta ray. An eighth grader in Sarasota, I am heading south along the sandy, outer shore of Lido Key on a clear day in 1946 just after the war, probably looking for whelks washed up by a storm. For some reason I stop and look over the Gulf. A good ways out, a huge, black manta ray soars out of the water, hangs above the horizon, knifes back in. Leaving me stunned in my tracks on the beach. I've never seen a manta before, but I know what it is; the name is instantly on my lips. This moment is given me out of the waves, suggesting a submerged world I know nothing about. Abruptly, my horizon cracks open, expanding not only westward but down into the Gulf. Unimagined possibilities add a new dimension to my life. I have no idea what they are, but now I know they are there.

3-2. Sunflowers. My partner lives in an apartment above her pottery studio. Bed and computer on one end, counter, stove and refrigerator on the other; a one-room apartment with no secrets because you can see the whole space and everything in it from wherever you are. I go up to get my camera case and come back down so we can go for a walk. "Do you like the sunflowers?" she asks. "What sunflowers?" "In the vase on the counter." I'd walked within six inches of them and never saw them. Not once but twice (*Consciousness: The Book*, p. 8).

3-3. Clouds up ahead. Clouds, nothing but clouds. I am looking for a first sight of the Rocky Mountains, but all I see through the windshield is clouds. Flanked by my brothers and two dogs, I am in the back seat of the car. My parents are in front. I am leaning forward, looking down the road toward the western horizon. Which is hidden by clouds. The family is moving to Seattle. We've driven from central New York to eastern Colorado, which is flat, offering long views ahead. Of clouds. I keep looking. Ten minutes. Fifteen. Half an hour. Nothing but clouds. I am about to burst with disappointment when, suddenly the white clouds, those same ones I've been peering at all the while—turn into snow-capped mountains. The Rockies! I see them! Nobody says a word. They've seen them all along (*Consciousness: The Book*, p. 10).

3-4. Rowing home. The McCormicks offer me a bed for the night, but I decide to row home. It's ten o'clock; we've just had Thanksgiving dinner. They didn't want me to eat alone, so invited me to their home on Butler Point. I accepted, but as I arrive by rowing across from the island, it begins to snow. Too late to change plans, I trudge up to the house through mounting drifts. By the time I leave four hours later, the storm has developed into a classic nor'easter. I head my thirteen-foot peapod into the lee of the point, then set course for the island. I feel the sting of icy snow against my right cheek, so use that as my compass to row south-southwest. With the wind almost astern, I make good time in total darkness. No stars, no lights, no traffic sounds, it's me, the waves, and that wind. In fifteen minutes, abruptly, the wind drops. I know exactly where I am: in the shadow of the outcrop at the head of the island. I hug the lee shore for half a mile, then haul up my boat and make my way to the cabin by flashlight. What a great dinner.

3-5. Atomic cannon. Bivouacking on a military training exercise in Western Germany, I spend several nights sleeping in a sink in one of the darkroom trailers my signal photo company (the only one in Europe as of 1956) brings along. I don't remember anything about the exercise except lining up for chow, that and fitting myself into my stainless-steel bunk at the end of the day. But on the return trip back to base, we run into a traffic jam in the center of a picturesque town with narrow streets. Inching ahead, we eventually come to the road-block: a giant, mobile artillery piece mounted between cabs at both ends is stuck trying to round a sharp corner while coming against us. It is way bigger than the cannon Ringling uses in the big tent to shoot a clown into a net—the barrel of the stalled vehicle looks to be over two-feet across. I know what it is: an atomic cannon. Here I am innocently sleeping in my sink and getting three meals a day, while the exercise is all about deploying tactical nuclear weapons against unseen forces to the East. I have a walk-on part in a much greater drama than I thought. Great powers toy with nuclear holocaust while the rest of us sleep.

3-6. Which way the wind blows. The standard joke when I arrive at MIT in 1950 is that you can tell which way the wind is blowing by the smell in the air: pickles from the Heinz factory to the southwest; chocolate from New England Confectionery Company to the northwest; the heavy scent of rubber from Cambridge Rubber Supply (?) to the north. But what really gets my attention is the contrast between the mechanical engineering laboratory in Building 10 (a room full of steam engines with ponderous fly-wheels and spinning brass governors) and the nuclear reactor out back that appears to have no moving parts. The first working transistor was built at Bell Labs in 1947, making vacuum tubes obsolete, sparking the electronic revolution. The winds of technological change are no joke; we can only lean into them as they blow from such centers as MIT and Bell Labs.

3-7. Being there. When we are alert to our surroundings so that our bodies and minds are both engaged in the same place, our consciousness comes together at one focal point, giving us a chance to make a suitable response to the situation we find ourselves in. I call that "being there," putting our bodies, attention, and values in the same place. A situation that is increasingly rare, given the popularity of cellphones, tablets, computer games, and the virtual world of digital media in general which treat body and mind as separable entities. I can't say that separating mind and body is necessarily a bad thing—videos shot in particular locales can quickly spread around the globe courtesy of platforms such as YouTube and Facebook, giving every videographer a chance to serve as the eyes—if not the soul—of the world. At the same time keeping the virtual world at a distance because the audience can't reach out and touch it in an effective way. When attention and action are divorced, we now call that multitasking, our modern way of having our cake and eating it too. But a physical body in a virtual world does not close a loop of engagement. We are at odds with ourselves, mind in one place, body in another. Being there, to me, means getting our subjective act together so mind and body are all of a piece.

3-8. Authenticity. Martin Heidegger made a distinction between being there of your own volition, and being there because you're following the crowd—doing the popular thing. He would say in the first case you are there *authentically,* whereas in the second you are there *in*authentically, not on your own but as a follower.

Consider the experience of Apollo 14 lunar module pilot Edgar Mitchell while returning from the moon in 1971. The spacecraft was rotating every two minutes to even-out heating due to solar radiation. The stars were brilliant against a black sky, and in succession on each rotation Mitchell saw Earth, moon, and sun appear in the window—rising and setting for him alone.

> It was overwhelmingly magnificent [he recalled]. I realized that the molecules of my body and the molecules of the spacecraft had been manufactured in an ancient generation of stars. It wasn't just intellectual knowledge—it was a subjective visceral experience accompanied by ecstasy—a transformational experience (Quoted by Richard Schiffman, "Do we need spaceflight for the perspective?", *Christian Science Monitor*, Jan. 23, 2012, from an interview in *Ascent Magazine.*).

Few people have been in a position to share Mitchell's experience on that mission. I would say that every aspect of his being was affected by the scene he witnessed from his privileged point of view so that, yes, I believe he was authentically *there* in the fullness of his being, making himself present to the universe on behalf of the rest of us. He went on to found the Institute of Noetic Sciences, an organization devoted to exploring the interface between consciousness and its setting in the physical world, a fitting response to the sensory impressions he gleaned on his foray into space on Apollo 14.

3-9. Voyager looks back. Another example of *being there* from an astronomical perspective is Carl Sagan's response to a photographic image of the Earth made by Voyager 1 on February 14, 1990, in which Earth shows up as a pale blue dot occupying 0.12 pixel. Adopting Voyager's perspective of looking back upon Earth across 3.7 billion miles from the outer edge of the solar system, Sagan combines that view with what it represents to his conscious mind *as if* he were on Earth and a passenger on Voyager itself at the same time:

> From this distant vantage point, the Earth might not seem of any particular interest. But for us, it's different. Consider again that dot. That's here. That's home. That's us. On it everyone you love, everyone you know, everyone you ever heard of, every human being who ever was, lived out their lives. The aggregate of our joy and suffering, thousands of confident religions, ideologies, and economic doctrines, every hunter and forager, every hero and coward, every creator and destroyer of civilization, every king and peasant, every young couple in love, every mother and father, hopeful child, inventor and explorer, every teacher of morals, every corrupt politician, every "superstar," every "supreme leader," every saint and sinner in the history of our species lived there — on a mote of dust suspended in a sunbeam.

Then, after continuing in the same vein, he makes his point:

> It has been said that astronomy is a humbling and character-building experience. There is perhaps no better demonstration of the folly of human conceits than this distant image of our tiny world. To me, it underscores our responsibility to deal more kindly with one another and to preserve and cherish the pale blue dot, the only home we've ever known (Carl Sagan, *Pale*

Blue Dot: A Vision of the Human Future in Space, 1997, http://creativecommons.org/licenses/by-sa/3.0/).

Here Sagan is speaking from two perspectives at once, one Earthbound, the other balanced on the edge of the solar system. Every writer of fiction performs the same magic in projecting thoughts into the mind of her narrator as if they were not her own thoughts. As Shakespeare created seemingly autonomous characters who were truly extensions of his own subjective understanding. As God him- or herself is claimed to serve as everyone's invisible friend. This is a trick our minds perform for us in being situated in one place or another, allowing us to look down upon ourselves as if we were in two places at once. If this gives us trouble when we don't like what we see, or invites us to give personal names to our separate selves, doctors will tell us we are suffering from a case of alienated self-consciousness, and can be cured if we learn to integrate our dissociated points of view.

In Sagan's case, where is he situated, onboard Voyager 1 or safely back home on Earth? Neither. His vantage point is firmly grounded in his own mind, and goes wherever it takes him. As an imaginative astronomer and astrophysicist, he has developed facility in traveling about the universe, taking different perspectives as he switches stellar or planetary atmospheres, *being there* wherever he projects himself through his gift of creative imagination, a further development of his personal understanding of the universe, much as a fiction writer projects herself by means of her hard-won understanding of human nature.

Being there is always subject to our personal points of view. That is, is a function of how we direct our attention. We can aim at a particular place or era in history, another person's mind, a roving spacecraft, or a different universe altogether. Even if we make up our travels as we go along, our thoughts are rooted in our understanding of our personal life experience, and we build out from there. It is that *rootedness* that determines whether, to use Heidegger's terms, we are being authentic or inauthentic in how we approach our own experience in one setting or another.

3-10. Being present. In the examples I have given, in contemplating the blue dot image provided by Voyager 1, Carl Sagan was doubly situated in his personal experience and his professional experience as a cosmologist, much as Edgar Mitchell was situated in both his sense of mysticism and his professional self as a highly-trained astronaut. I would say both are authentic in being themselves in ways that integrate large portions of their respective life experience. If they are along for the ride, they participate fully in whatever events they encounter, and engage their sensory impressions with an acuity earned through the care with which they extend themselves—put themselves out—in opening-to and embracing their new surroundings.

In that sense, both Mitchell (*see* Reflection 3-8 *above*) and Sagan (*see* Reflection 3-9 *above*) are worth listening to because each is wholly authentic. As opposed to me in the incident of my seeing the manta ray (*see* Reflection 3-1 *above*), where I was in the right place at the right time to witness an extraordinary event, but had no sense of what it might portend because I was present by accident during the inauthentic, early days of my apprenticeship as an observer. Similarly, in forcing snow on the Rockies to do duty as clouds (*see* Reflection 3-3 *above*), I was projecting my inexperience with snow in late August because I was accustomed to first snows arriving in November, so had difficulty being where I was in briefly passing through eastern Colorado. In not seeing my partner's vase of sunflowers (*see* Reflection 3-2 *above*), I was in such a hurry and so intent on retrieving my camera case that I had eyes for nothing else, so was not present to the sunflowers, and they were not present to me. In both incidents of my rowing through the storm (*see* Reflection 3-4 *above*) and confronting the atomic cannon (*see* Reflection 3-5 *above*), I was in the *process* of learning how to be fully authentic in novel situations, but had little experience in reaching beyond myself, so could feel my awareness growing as I incorporated new episodes into my expanding repertory of life experiences. What I learned from my two years at MIT (*see* Reflection 3-6 *above*) was more a function of what I was ready to pick up on my own than what my teachers had to convey. Which proved equally true during my subsequent two years at Columbia in what, for me, was a novel and inviting adventure in urban education.

3-11. Arousal. Consciousness is a state of heightened arousal in which perception and action are joined in a seamless loop. When our eyes make random motions while we dream, we are largely awake; it's just that we can't create sensory impressions from ambient energy impinging on our sense organs, nor can we will ourselves to act in an appropriate manner. Instead of being awake to our surroundings, we are awake to whatever drama our minds concoct from the fact that we cannot engage, often leading (in my case) to dreams of frustration, which is what I feel when I can't interact. Or once engaged, get stymied or interrupted. Dreams are a way station between sleep and full wakefulness, a kind of warm-up in the bullpen before being sent out to the mound aroused, awake, and fully conscious.

Sleep, blessed sleep. It readies us for the promised adventure of a new day. We don't yet know what challenges we will face, but we do know we will be tested in new ways, so must rest while we can. Then comes the alarm bell, a blast from the clock radio, scrape and crash of garbage cans thrown onto pavement, horns honking, sirens wailing, crows cawing, brakes squealing, voices calling, dogs barking—and the day begins. One way or another, we get the message. Not just from the room getting lighter and noisier, but from our midbrain reticular formation—our inner klaxon—calling all neurons to assemble on deck in a hurry.

We rush through the standard routine: somehow feet get put on the floor, eyes popped open, heads scratched, clothes put on, breakfast downed, teeth brushed, forecasts consulted, clocks checked, doors opened—and we're off and running for the bus like a herd of windup toys all set down at the same time to head off in every direction. We devote our life's energy to whatever it is we do all day, then return to our starting place, strive to be civil to one another, start to wind down, do what we can to relax, make our kind of love, and do our best to wear ourselves out so to be ready for the klaxon to sound in the morning.

It is startling to consider how much energy we put into maintaining our daily routines, cumulatively adding to the lives we lead between wakefulness and sleep. I picture those routines as made up of multiple parallel skills and capacities linked together in serial order to form a continuous loop. Arousal enables memory, memory enables expectancy, expectancy enables attention, attention enables sensory impressions, impressions enable recognition, recognition enables understanding, understanding enables creative imagination, imagination enables meaningful situations, situations enable judgments, judgments enable decisions, decisions enable the setting of goals, goals enable projects and relationships, and these in turn enable actions in the world. Which come back to us in relation to our charged expectancies, hopes, and desires. No wonder we exhaust ourselves by engaging in our everyday activities, and fall into bed to get ready for picking up the thread of our engagements the next day. *Being there* hinges on our being aroused so we can get out of bed to salute the sun—or its electrical step-child.

3-12. Expectancy. We hold every hope of waking to the same world we went to bed in last night. But in spite of our wishes, Earth keeps turning on its axis all night: somewhere cyclones are reaching land, volcanoes are fuming, earthquakes rattling dishes, houses burning, babies being born, people dying. Which means we might have to reinvent ourselves while reading the morning paper or watching the news. We may put ourselves on hold while we sleep, but the universe we are a part of never closes its eyes or takes a day off, so change is inevitable, both in our surroundings and in our most intimate selves. Expectancy is a reflection of what we are used to, but the world we are used to may have burned overnight and a new one risen from the ashes.

Expectancy is rooted in habit and memory. A sense of probability based on prior experience, it is what, all things considered, we think will be likely to happen. Too, wishful thinking skews our outlook, so expectancy sometimes reflects more what we hope will happen as an improvement over our actual experience. Or, again, dread may color our thinking, so what we expect to happen will prove to fall short of the past. In the days of blind dates, who did not become jaded in setting up the next one, perhaps vowing to abandon the date market altogether. The point being that our view of the future is governed by trends set in the past, so we learn to expect the future to fit the envelope defined by the variety of our earlier experiences.

Without habits or memory, we would have no basis for expectancy one way or another. Expectancy sets the stage for comparison of the now with the then, and a disparity in that comparison—for better or for worse—is what alerts consciousness to attend to the difference. Arousal, memory, expectancy, attention—all are essential to consciousness of sensory phenomena in the mind's effort to answer the question, *What's happening now?*

3-13. Attention. We don't confront sensory phenomena until we pay attention to something that captures our notice either outwardly or inwardly. In an inward sense, being hungry has a kind of feel to it, sometimes a gurgling or growling sound that gets our attention because it is biologically salient. Being thirsty is a dryness of lips, mouth, or throat. Thirst and hunger are motivators that set us working toward our next meal; without them we'd not be long for this world. Anything to do with sex is a real attention grabber, starting with an inner urge and working outward from there. Being tired shows up as having trouble paying attention, so we pay just enough as a reminder to shut down for the night.

We all have special interest in and concern for our current situation in this world, so pay special attention to sensory clues that update us about what might be happening (if we interpret the patterns correctly) in the scene around us. Some patterns are inherently salient: colorful sunsets, starry nights, sun and moon, storms, moving animals, projectiles hurtling toward us, thunder, bright colors, sharp objects, babies, smiling faces, clear complexions, people walking toward us, people running away, and on and on. Other patterns are salient if they conform to our interests, concerns, expectancies, and lifestyles—hobbies, collectibles, favored athletic teams, playmates for our children, workers we might hire to do a job, doctors and nurses to watch over our health, civil servants who help us feel safe . . . again, on and on.

Attention invites such salient stimuli from without or within to enter consciousness one at a time so we can incorporate them effectively into our ongoing engagements. We trust them to be there when we need them, and if they're not, we direct our attention toward finding out how we can reach them. If we can't put a name to a face that we recognize coming toward us on the street, we go into search mode by running through the alphabet, thinking what the absent name might sound like, or trying to recall the setting where we last saw that face as familiar and attached to a name. Nothing is more important to our affairs than controlling how we direct our attention in a useful manner. The world is a huge place and we can only address its complexity one detail at a time, putting the rest on hold for the moment.

3-14. Acuity. We can simplify our sense of the world by dealing with it in large chunks rather than focusing on each telling detail one-at-a-time. We can do this conceptually by abstracting the gist (say, the emotionally-charged high points) from the full range of our experience, or by simply leaving out what is inconvenient to deal with under current conditions. In daily practice, we employ a sense of scale in approaching sensory impressions, choosing to regard only the details we are interested in at a particular level of discernment or discrimination, saving the rest for consideration at another time. Regarding a rocky landscape, we can parse it on the level of outcroppings of bedrock, boulders, cobbles, pebbles, gravel, sand, or particles of clay, scaling our scrutiny to match our level of concern. When children ask where babies come from, we can scale our answer somewhere between a full biological explanation and a complete dismissal of the issue, probably settling for an amount of detail tailored to the specific occasion (from a child's point of view) on which the topic comes up.

Conceptualization is a useful way of grouping similar phenomena according to a small number of defining charac- teristics, saving time and mental effort by skipping over indi- vidual particularities. For example, putting aside their botanical and perceptual differences, when convenient, apples and or- anges can be lumped as two kinds of fruit. But when individual differences are the essence, as when trying to deal equally and fairly with a widely divergent population of distinct persons of various ages, education levels, races, language groups, sexual preferences, religious beliefs, income, and so on, dealing in conceptual chunks is apt to do violence to some members more than others because of disagreement about what might consti- tute a "normal" person. Now that corporations are legally regarded as persons, confusion and inequality are built into any notion of what a person might be.

3-15. Recognition. My experience with ambiguous figures (which may not be ambiguous so much as figures I see incor- rectly) has taught me that things are not always what they seem. My friend Fred seen from the back on Fifth Avenue—he had Fred's raglan-sleeve coat, Fred's cordovan shoes, Fred's scarf, even his gait—turned out, when I finally caught up with him, to be a stranger passing himself off as Fred in full public view. I have seen a dying crow by the side of the road morph, once I got close enough, into a black plastic bag blowing in the wash of speeding cars. The crashing jet I glimpsed over the roofs was really a swept-back TV antenna in Bar Harbor. The man scraping paint off the side of a house in mid-winter was a man-sized, northern white cedar tree blowing in the wind. The most complex—and durable—ambiguous figure I have ever witnessed was a cartwheel display of aurora borealis:

> The center of the spectacle is at the zenith overhead, apparently directly above my island camp in Franklin, Maine. It is midnight. On my way back from the latrine, I look up—to see streamers shimmering from around the horizon toward that celestial vortex where, wavering, flowing, they whirl together in a pulsing gyre of living forms that spreads and contracts and shifts its shape as I watch. Glowing spiders turn into snakes into eyes into butterflies. The air is clear, sky dark. Each star is a vivid mote of light. . . . I am having a whole-body experience. Candle flames turn into running wolves into great whales into chickens, rays shooting above the trees all the while, feeding the starved gyre, spinning it round and round and into itself. Roses turn to sparklers turn to ants turn to luminous lizards. The spectacle goes on for hours, each second consuming my entire attention. What if I blinked and missed a crucial transition? Continuity is of the essence. But eventually, cold, stiff, tired, I not only blink but go to bed, my head swimming with the best auroral display I've ever seen—and am likely to see in my lifetime (*Consciousness: The Book*, p. 17.)

I knew in my mind that the light display I was watching was generated by charged particles radiating outward from the sun, then interacting with Earth's magnetic field. But what I "saw" was such figures as I mention in the above quote. Solar energy may have given rise to the aurora, but it was my lived experience that fleshed out that energy as the objects I "recognized" in those swirling lights. *In Consciousness: The Book*, I labeled such apparitions as "category errors"—mistakes made by sorting sensory impressions into the wrong conceptual bins. But they aren't really errors: they are what I made of what I saw at the time. I offer such incidents here as evidence that what we see (hear, touch, taste, etc.) before us is not a given but a personal judgment in which we participate. Whatever category makes the best fit to the phenomena passing through our minds is for us to say and is not foreordained by any panel of experts. Philosophers say consciousness is *intentional* in always being *of* one objective thing or another, but what a thing *is* is a matter of convention, not a revealed truth. When is a stream a brook, a creek, a run, a rill, a branch, a crick? Who's to say? Exactly! The one who calls it by the conventional name for such a phenomenon as based on personal experience in his or her language community, that's who's to say, and none other—or rather, *all* others who have a favored name for such a thing. If it looks like a stream, and sounds like a stream, and flows like a stream, I calls it a stream because, to me, that's what it is. If you speak a different language or dialect, I suspect you'd be calling it something else. Which gives recognition of what a thing *is* a kind of metaphorical ring—as I named those streaming lights after what they spontaneously reminded me *of,* not what they "were." Right or wrong, as we recognize a sensory impression, so do we call it, whatever name best applies in those circumstances.

My point is that how we see a thing and what we call it are matters of convention within one cultural community or another, not some kind of inherent identity. Which, then, comes first, being or naming—what a thing *is,* or what it is *called?* Each of us must decide such questions for him- or herself, as she feels supported by members of the family, neighborhood, and community that taught her *how to be* and *how to name* in the first place. That is, taught each of us how to recognize ourselves and our actions as *familiar* (of the family) on different occasions, so we'd know who we are and how we ought to behave at all times. When we get up in the morning, we assume the world is still the same, old, recognizable world we went to sleep in—until we are disabused and have to reinvent ourselves yet again. By definition, those with a conservative bent have difficulty updating the world to keep up with changing times, so put every effort into forcing events to fit the mold they have grown accustomed to, even if largely a fantasy. Meanwhile, more adventurous types do all they can to move a staid world away from its respectable traditions in achieving the glorious (and equally fanciful) vision they see as their appointed destiny. So, for example, might they march stalwartly ahead with their cap visors shading their necks, not their eyes.

Where I use *recognition,* others might use *categorization* in referring sensory phenomena to concepts already in place in their memories, together with familiar labels for particular chunks of salient experience. The overall principle is *interpreting* what a given thing *is* to us in our personal experience. Not only deciding what to call it, but judging what role it is to play in our lives—at least for now. How an impression strikes us—how we take it in—is a function of our outlook, our perspective within a particular situation. And that perspective, in turn, depends on our history of personal experience within the life we have led up to now.

"Divorce," to take one example, has a different ring to it before we undergo such an experience than after we have actually lived through one or more. The abstract, conceptual category referred to by that name has grown far more specific than it was, each detail now burning with a particular heat that it lacked in earlier, more innocent days. The concept itself has acquired specificity in terms of our personal experience, leading us to revise our outlook on ourselves and the world for better or for worse. We now recognize divorce in a new light because it has taken particular shape and coloring through our intimate participation in it, and it has become an integral and a vital part of our autobiographical memory, which is the personal history of our stay on Earth as we have actually lived it day-by-day. That is, as we have experienced it on the inside, not as any so-called impartial witness might have observed it from an outside, supposedly objective, point of view.

Beware objectivity. It rules out the personal colorations of subjectivity, so strips conscious life of its inherent vitality. *Categorization* is a philosopher's word for the profoundly personal experience of reaching out to the world as unique individuals, which in every case is the best we can do. I prefer *recognition* or *interpretation* for sorting sensory impressions into types because such terms at least acknowledge my central role in living my own life. Self-reflection, the language of subjective consciousness, flows inside-out, so sometimes requires a different language than impersonal observation meant to appeal to consensual agreement. Introspectors can only deal with life as it is lived by one subject at a time; philosophers, theologians, and scientists are generalists who impose conceptual models on the whole of humanity at the expense of each member's demonstrable uniqueness.

My view is that, in recognizing sensory impressions as fitting one labeled group of chunked experiences better than another, we are more freely and accurately conscious than if we impose standard definitions describing how things are to be chunked according to an intellectual model of what consciousness ought to be, thereby disposing of what I take to be the essence of personal consciousness—its origin in and adaptability to personal experience. My world is precisely as I have lived it, conceptualized it, and now take it to be, warts and all. Those are my warts. Without them I wouldn't be me. Yes, I can have those warts removed, becoming a new me. I view perception to be more metaphorical than literal in the sense that it arises in the awareness of a unique individual, and so necessarily reflects the idiosyncratic outlook of that singular person, not of his language community or any prescriptive dictionary that matches words to their authorized definitions. Introspection *describes* specific instances of personal experience in the face of cognitive models that *prescribe* in general terms what experience ought rightfully to be.

3-16. Time, space, and attachment. As you have gathered by now, my consciousness is my own, and my words flow from that unique source. I am not so much writing about consciousness in general as living it in a very particular manner as I write these words, trying to match what I say to my ever streaming thoughts as I sit here and now, fingers poised above the keyboard of my computer, writing, deleting, changing, building paragraph after paragraph, trying to remain true to perceptual experience as a fundamental feature of my mental life. What is it like to live in a flow of sensory impressions? That is the question I am dealing with, and you are dealing with in your own way in reading these sentences, translating them into your own stream of awareness as you go.

What is evident by now is that I am attached to my stream of consciousness because I am figuratively swimming in it as the very medium of my life, my awareness, and the person I am, while the same is true of you swimming in the very different stream of life awareness that makes you who you are, with resulting attachments of your own. We cannot be conscious without becoming attached to our particular version of what we sense, remember, think, feel, and do. This is our life; what we make of ourselves; who we are in this world.

In my case, I am a highly visual person who makes images photographically, largely focused on my natural environment. I gather those pictures into slide shows and PowerPoint presentations. I cook my own meals. I think my own thoughts and write my own books based on personal experience, which I illustrate with photographs I have made. I do not paint, sculpt, make music, or dance. I do not play football, baseball, basketball, tennis. I have done such things in the past, but don't do them now. I have two sons, one a glass-blower, the other a chef. My partner of nineteen years is a potter. I speak English, not Mandarin or Arabic.

For almost thirty-five years, one of my main concerns has been to account for the fact that each person I have met has a distinctive take—both generally and specifically—on life, personal affairs, and the universe. To that end, I reflect on the only specimen of consciousness available to me—namely my own—to keep track of where I am coming from in forming the impressions, thinking the thoughts, and performing the actions I do.

My biggest discoveries have been that I create my own time and my own space in the process of conducting my affairs, and am deeply, irredeemably attached to this life that I lead. If I did anything differently, I wouldn't live in *my* time or *my* space, and I wouldn't be me.

Due to my own bodily motions, space, as I now view it, is the product of my unconsciously deleting changes *to* my sensory impressions *from* the confusing picture I would otherwise see as a composite representing both changes in my perspective and changes in my environment as I go. My intent is to form an accurate picture of the terrain through which I travel and act, and of the placement of objects in that space *as if* I had not moved at all. Strange business. Yet we all do it in keeping our wits about us, so to stay on an even keel as we move. The resulting picture we create for ourselves is the landscape through which we move purified of the distortions we would otherwise introduce by moving place-to-place and continually shifting our attention and point of view.

Our brains make these adjustments for us automatically so we don't have to think about them. But birds, I believe, aren't so lucky. Watch a robin standing still on the lawn, then abruptly shifting its neck to a new position, giving itself time to update its relationship to its surroundings in order to make a deliberate lunge at the worm it senses beneath the grass ahead. Observe the jerky movements of crows and starlings, which I suggest might accomplish the same goal of shifting from one clear scene to another as a series of snapshots without having to deal with the intervening blur in between. Soaring, too, makes it easier for birds to map the world below by not having to concentrate on flapping all the time, as does hunting while sitting still on a branch as eagles do. The alternating saccades (quick shifts of attention) and points of fixation (and clear vision) in our own eyes achieve the same result whether we are reading a book or scanning an image; the shift happens so fast we don't lose our place in grasping the whole, sparing ourselves a fit of confusion or vertigo.

Time, too, is of our own making as a timescape available to us when we stand or sit still in a seat and let the world change around us while we make only minor adjustments to the focus of our attention. Marksmen and snipers do the same, steadying themselves so to fix the crosshairs in their sights on the far target they are aiming at, bringing the two into alignment, pulling the trigger at the right instant, dividing time into hits and misses. Listening to music, watching a game from the stands, making ourselves present to a theatrical performance, we divide our experience into units of time—beats, melodies, downs, innings, scenes, acts—while we sit still and enjoy the changes taking place before us as they develop yet hold our attention. A string quartet is a repeatable timescape we can venture into, as is a song, jazz performance, chorale, or symphony, or any game we can vicariously participate in from a seat in the bleachers. Written out as a score, music takes on a spatial appearance, like the pages of a book, but each bar is played in serial order, as each word, sentence, and paragraph is read, producing a flow of experience matched to the streaming nature of our consciousness, producing a coherent performance achieved as a timescape in personal awareness.

Whether exploring spacescapes or timescapes, we are the same person, moving on one occasion, remaining still on the other, giving rise, once we calibrate them in meaningful units, to an active sense of space around us and a more passive sense of time (in that we are not responsible for the coherence of changes taking place in our awareness, so we can sit back and relax). We not only witness these *self-changes* and *it-changes* (*see* Reflection 1-18 *above*), but become attached to them as memorable experiences built into our lives place-to-place, minute-by-minute, adding to nothing less than our earthly experience, to which we become attached because it is uniquely ours—respectively mine and yours. If we go blind, we can listen to music, broadcasts, and talking books, maintaining the dynamic play of our conscious lives. If we become deaf, we can take a new look at the world around us, walking, reading, collecting stamps, doing picture puzzles, playing croquet, going to museums, being where we are in life in new ways, discovering new attachments to the persons and events that sustain us in the process.

What I am getting at is the engagements we use to consciously or unconsciously connect our sensory impressions to the actions we perform in leading the lives that we do. To the degree we are in control of those engagements, we have a say in determining how to navigate through the respective journeys we have ahead of us. These engagements serve as our connections to the world we live in, suiting us to be where we are, when we are there. *Being there* is not a matter of placing us down on a board like so many chess pieces, but more of our discovering who we are, where we are, when we are there, creating powerful attachments to our ways, our situations, and our surroundings in the process because we are engaged with them as the basic elements of our lives.

Consciousness, then, serves as the medium in which we exist. From a phenomenological point of view, that is exactly the case, in spite of the popular belief that we, like so many vibrating molecules, are inhabitants of a physical universe we cannot explain, yet have concocted around us in order to feel comfortable with ourselves. Or a spiritual universe, for the same reason, with similar results.

So here we are, like Edgar Mitchell, riding in our mental capsule rotating through time and space, being transformed in the process, looking out on stars, sun, moon, and Earth, doing our best to understand it all, while time, space, and our attachments are wholly our doing. Beyond our understanding of things there is nothing about *being there* we can hope to understand in any way better than by being wholly ourselves in living out the destiny that *being here* on the inside of our minds makes possible. *Being* either *here* or *there* is a gift of engagement from the universe that combines matter and energy within us to produce consciousness. Each of us is an instance of that feat, far more wonderful than any mere dream of turning a lump of gray lead into gold.

3-17. Information processing. I do not rely on the trio of concepts that has become standard in the world of cognitive neuroscience (information processing, computation, knowledge) because I have not met any of its members in reflecting on my own mind. *Information* requires a proper mental frame of reference if it is to be interpreted rightly, and I am never sure if the energy I receive from my physical environment ever ends up in the proper conceptual frame or not. Even if I were calibrated in one discipline or another to fit neural signals to such a frame, I could not be sure I'd gotten it right because I'd be basing my interpretation on someone else's judgment, not my own. Science has its successes, but I know just enough about it (as I know just enough about wild mushrooms to avoid eating any of them) to recall all the times that such calibration is overturned by yet another paradigm shift or new way of thinking about the mind and its brain. I feel more comfortable not going along with the current trend of assuming neural signals are meaningful in and of themselves.

As for information processing as a form of *computation,* I'm not comfortable with that, either, because nothing about my mind and its brain reminds me of a computer—though I recognize that such a metaphor has broad appeal to a great many others. When I briefly worked as an engineering aide at Boeing before being drafted in 1955, I used a slide rule that added or subtracted logarithmically and a clunky mechanical calculator with a handle you had to pull down like on a slot machine of that era. Others used an abacus. I've never caught up with the idea of seeing the brain as a computer because it seems to me that the attractiveness of machines (because we have to understand them in order to build them) does not apply to the brain as the most complicated organic system in the known universe, a system we didn't design or assemble, and don't understand.

As for *knowledge,* I am leery of the term because commonly one man's certain knowledge is another's fashionable jargon. I use "opinion," "belief," or "interpretation" to give notice that my experience convinces me I may be right, but I am unable to demonstrate the truth or universality of what I believe. Science runs on consensus and repeatability of results; I am only one out of seven billion (often conflicted) minds, so am hesitant to rise above my modest station in life by unduly asserting my views. If I seem to be doing just that in these pages, that is because in speaking of "the" mind, I am writing about a sample of one.

3-18. Cultural consciousness II. We are born to a world offering many ways of being conscious: the Moses way, Buddha way, Jesus way, Mohammed way, Socrates way, Machiavelli way, Gandhi way, Shakespeare way, Stravinsky way, Picasso way, hunter way, farmer way, builder way, shopper way, warrior way, family way, teacher way, and all the other ways of the world into which we are thrown, and pick up in the process of growing into ourselves by *being there* with eyes and ears open to what's happening in our familiar surroundings.

We learn to be conscious by modeling ourselves on (or rebelling against) the actions of others, though we can never be sure of the perceptions and situations that precede them. That is, the connectivity of our neural networks is shaped by those who accompany and nurture us when we are young, providing us not only with food, warmth, and shelter, but with outlooks, attitudes, expectations, and abilities as well. Very little of the process by which we become conscious is understood, much less scripted. It just happens, as the Zeppelin airship Hindenburg just happened to catch fire during the last minutes of its arrival from Germany at Lakehurst Naval Air Station in early May, 1937. We still don't know what started the fire, but we do know that 36 people died (62 survived) as a result, and the crash put an end to passenger transport by so-called lighter-than-air craft.

Similarly, I don't know why I am conscious as I am, but I can hazard a list of probable influences: my grandmother's death on the day she gave birth to my father; my parents' upbringing; the order of my birth among three male children; the times to which I was born and the place, Hamilton, New York; John B. Watson's method of child rearing as promoted in his book,

Psychological Care of Infant and Child (1928); among many other factors affecting how I reach for sensory impressions to this day, how I am situated in my own mind, and how I am led to act in the world. Such factors have affected the particular connectivity of my neural network, and therefore the underpinnings and neural correlates of my singular consciousness.

In a very real sense, the cultures in which we participate serve as correlates of our subjective minds, raising the question, is this my thought, or does it belong to my situated presence on Earth? Some thoughts are not possible in particular surroundings, but might well become current twenty years later, somewhere else. And conversely, my own sons won't be able to share some of the key thoughts I have had in my time and place. What is true for one age might well be considered a damned lie for the next. Since each of our minds is unique, the cultural correlates of consciousness can never be known in fine detail. You had to have been there to get the full picture, and now it's too late.

3-19. Levels of consciousness II. Since the point of both perception and consciousness is the same in leading to effective and appropriate action in the world, sensory impressions do not stand alone but are a way station in our ongoing engagements with the world around us. There are, however, two main routes leading forward to action: a relatively fast and direct route, and a slower and more challenging—and, I would say, more scenic and rewarding—route via what I call the situated self at the heart of conscious deliberation.

Having come this far, *being there* has brought me to the fork where the two routes diverge. We either grasp our situation *unconsciously* well enough to proceed directly to action—or we need to reflect on our options in order to make a *conscious* and well-informed judgment about what next steps to take, so settle on a mental detour around direct action to explore the situation we find ourselves in, while giving play to our values, feelings, assumptions, understanding, and creative imagination before deciding what to do.

In the first case, moving unconsciously ahead makes sense from a variety of different perspectives, all amounting to the fact that we already know what we should do, so advance directly from perception to action without further delay, as in the following examples:

1. Our response is a *reflex* we don't have to think about; we just do it automatically.
2. We've seen others do the same thing, so simply *mimic* them.
3. We learned that in school, the answer is ours by *rote*. 7 x 8 = 56, what else would it be?
4. It's a meaningful *routine* we go through every week: Thursday means bowling, period; what's to decide?
5. We prefer to do the current thing, that's our personal style (or *bias* if you will). Being up-to-date, even if it seems like a foregone conclusion or a *prejudice*, that's who we are. We say, out with the old, in with the new.
6. It's how we believe it should be done because that's how we've been trained or brought-up. *Orthodoxy* is just being true to our deepest habits and beliefs.

These are all time-saving variations that get the job done quickly, more-or-less correctly, and efficiently—in the manner we became accustomed to a long time ago. The alternative is to resort to *full consciousness* by facing into our ambivalence about making a decision, starting from scratch by consulting our feelings, values, understanding, imagination, hopes, fears, and so on in coming to a sensible decision. That means going beyond what we've done up until now, requiring us to break new mental ground.

The way of full consciousness opens before us in unique situations we've never encountered before. Perhaps this is our first time being in this situation, and we've got to consider every detail, so it makes sense to go slow and figure things out as we go. Maybe we're new at this, so have to take time to get up to speed. Perhaps some of us are scientists and have to get it right to avoid being seen as upstarts and get put down by our colleagues. Writing a paper that's going to be peer-reviewed, that may be your definition of torture, but it happens to be how the game is played. If we want to be professional, we have to play by the rules. Or, perhaps we don't know the answer to a particular problem, so have to do our homework by figuring it out on our own. No, the answer is not in the back of the book; that's exactly what we've got to work out for ourselves.

As an example, I took my boat mooring up this morning, a muddy job, indeed. I've been doing it the same way for eight years, putting on my rain gear and boots, walking in the mud at dead low tide, undoing the stainless steel wire binding the shackle in place, undoing the shackle which has been underwater for eight months, bringing in the buoy and outhaul rope, letting rain wash mud off the rope, and so on. It's the messiest job of the year and I and my gear end up covered all over in sticky, gray-green mud. Each time I vow I'll never do it again—until I find myself in April putting my mooring in place and taking it up eight months later. Walking back from the grocery just now, I said there's got to be a mud-free way of setting and recovering that mooring. What if I did it at mid-tide when the mud is underwater and I am in my boat, reaching down with a gaff to hook the chain, bringing it up, attaching the buoy and shackle without having to wade through the mud? Hugging my grocery bag, I could picture the whole thing. Next year will be different, I told myself. And it will be—if I remember to put my plan into action when April next comes around. (*See* Reflection 1-9 *above.*) ○

Chapter Four

ACTION

4-1. Ceiling tiles. I still remember sitting in the back of the room in second grade, not being engaged in what we are "studying," looking up at the ceiling tiles overhead, wondering how many holes there are in each tile, so counting them again and again because I am always unsure of my total, and have to check to make sure I am right. I have standards at least, even for meaningless tasks. I am into counting, not yet multiplication. The only way to be sure of the answer is to repeat the process one more time—which always leads to a different answer, so I find myself consuming the minutes by counting into the hundreds, when I'm always interrupted and have to stop and pay attention to something else, so I never find the answer to the question those tiles pose to my wondering mind.

4-2. Prima donna. In college, my advisor has to approve the classes I sign up for in the next term. He is big on academic rigor as a builder of strong minds, so looks down on the arts. On the practical arts in particular. Classes where you'd learn to draw, or paint, or make movies. These are early days after World War II, and scholarly standards are staunchly defended against erosion, which is thought to have happened during the war when many of the best teachers served the nation, leaving draft-deferred substitutes in many classrooms. In my case, the class at issue is block printing, specifically color woodblock printing, as taught by Hans Mueller, a craftsman in the European tradition. The class isn't taught every term or even every year, but next term it is, and I am set on taking it. Which I do, against my advisor's better judgment. For which he calls me a "prima donna" for insisting on having my way against his will. A visual learner, I will remember that class all my life, where many more conceptual courses will be buried deep in the waste heap of my mind. I do learn to make woodblocks, and practice that art along with photography for many years.

4-3. The general gives a speech. After basic training at Fort Ord near Salinas, California, I go to the Signal Photo School at Fort Dix in New Jersey. As a draftee, I am lucky to get in because photo school is increasingly being reserved for enlistees who are in for three years and I as a draftee am in for only two. My Classification and Assignment officer takes sympathy on me when I tell him the only skill I have is photography, and I'd be a waste to the Army if I did anything else. Trained as a lawyer, he himself wants to be in the Judge Advocate Division, so does me a favor. My experience has been mainly with 35mm photography, whereas photography in the Army means shooting with a Speed Graphic using 4" x 5" cut film. I do the sixteen-week course in eight weeks, so don't have all the training with large cameras others in my class do.

I remember it well, one of my last assignments: Come back with a picture of the general making a speech. When I arrive, I find the general already speaking in a large auditorium, every seat filled with a soldier in fatigues. He stands behind the podium in center stage, so to avoid featuring the podium in my shot, I figure I'll get up on the stage with him and shoot from the side. It is too dim in the auditorium to trust my rangefinder, so I estimate the distance, put a flashbulb in the socket in the reflector, cock the shutter, insert a film holder, pull the slide, look through the viewfinder—and push the bulb-eject button, not the solenoid shutter-release. The bulb flies out of the reflector and bounces on the stage at the general's feet. He looks down at the bulb, then at me with a wry expression on his face, and goes on speaking. I hadn't wasted a sheet of film because the shutter didn't go off, so pick up the bulb, put it back in the socket, frame the general in the viewfinder, and push the right button. This time the bulb explodes because it cracked when it hit the floor, glass spraying at the general's feet, and I have wasted a sheet of film. I doggedly cock the shutter, turn the holder around, pull the slide, insert a new bulb, and on the third try, get my picture.

4-4. Sixty notebooks. From 1993 to 1998, I make sixty different hikes in Acadia National Park, scribbling the details of each hike in sixty different little spiral-bound notebooks hung by a deer-hide bootlace around my neck, stub of a pencil poked through the spiral. I divide the hikes by the four seasons, and make one hike a week for fifteen weeks, suiting each hike to prevailing trail conditions, sunlight, snowcover, precipitation, wind direction, and so on. I then write up those hikes from my collection of notebooks, and get ready for the next batch of fifteen. It takes me four years to make, illustrate, and write-up the sixty hikes, and another year to do the maps, select photographs, make an index, and have a proofreader check the manuscript I have laid out in WordPerfect. Hiking, writing, taking pictures, making maps, I am totally focused on my project. To me it is the good life, which I live on my own time while working as a part-time ranger in the park. For my efforts, I get delivery of a thousand copies of *Acadia: The Soul of a National Park,* a book I now see as being more about my soul as a hiker than the terrain where I hiked.

4-5. Long division. In sixth grade, I have trouble learning long division. My teacher runs through the steps in class, but I have no idea what she is doing, or I am doing, or why I am trying to do it. In my small life, it makes no sense at all. I like writing, but I don't like numbers. They are just something teachers make you do. I remember being kept after school one day to work on long division. It starts with a fraction, one number on top of another. You make this funny half-a-box, put the top number inside, the bottom number outside on the left, and write the answer (when you figure it out) on top of the box. The trick is to match the bottom number to the top number to see how many times it fits in. It's like multiplication, only backwards. You start on the left and work to the right, making a guess of how many times the bottom number fits in, writing the answer, then making another guess. If it doesn't fit, that's called a remainder. You draw lines under the box, multiply the bottom number by how many times you say it fits, then subtract the total from the number you've already got, and keep going until the bottom number won't fit any more times. That, together with any remainder you get, if you get one, is the answer.

What my teacher shows me is that it is just a series of steps you've got to go through—same steps every time. That is the trick to division, breaking it down to one step after another. If you remember to do all the steps, that's called long division. What she teaches me is to do the steps in the right order, no matter what numbers I start with. Even if the numbers aren't the same as last time, the steps are always the same, and the answer on top of the box is the right answer. Suddenly, I get it. Same way every time. Easy. It's not a problem if you know what to do. Long division is not a mystery, it's a series of things you do to get the right answer. If you can multiply and subtract, which I can, you can get the right answer out of the box. *I* can divide! That day after school changes my life. I go from being a dummy who always has a hard time with numbers to being a student who can learn (and subsequently forget) similar routines all the way up to solving differential equations.

Much later, electronic calculators get invented and they do the steps for you, so you don't really have to understand long division. When it comes to performing set routines, calculators are faster and smarter than I am. So I don't do long division much anymore. That is called progress.

4-6. Cello lessons. Learning to play the cello is like long division without right answers and without steps that I can remember. My mother thinks playing the cello would somehow make me a better person, or one she'd like better, but it almost kills me. I take lessons, I practice, I get hauled out for my mother's friends, but I just don't get it. It isn't taking lessons, it's learning to read music so I make the right notes by turning black marks on paper into something that sounds like music. When I get to ninth grade, there are three of us in school who "play" the cello, so we are all enlisted into the school orchestra. Faith (an eighth-grader) really can play the cello so it sounds like music. I sit behind Faith and try to do what she does, but everything goes by so fast I can't keep up, and no matter how I wave the bow or place my fingers, it never sounds like anything close to music. In performance at the Christmas concert, it is terrible having to pretend I can play the cello, scraping my bow back and forth without touching the strings, wiggling my fingers as I think Faith is wiggling hers. One of my brothers plays the piano, the other the trombone, I am the one in the middle who can't play anything. I hate myself, I hate the cello, I hate my mother. I end my musical career in ninth grade, and am a happier person ever after.

4-7. Fossils. At the same time I was miserable playing the cello, I was having a great time exploring streambeds cutting through the local shale, looking for fossils. Years earlier, my father had had a load of stones delivered to put on the driveway. Several cubic yards of small pebbles, it was like having a beach delivered to our house. I studied them closely, loving them for their heft, smoothness, colors, and designs. I remember selecting one I liked, taking it into the garage, holding it against a cement block, and hitting it with a hammer. Shazam! It split neatly in two, revealing a fossil shell I never suspected was there, and the hollow impression of that shell. Striking that stone at random for no reason changed my life. I've been in love with fossils ever since. Forty years later, when I was in grad school at Boston University, I wrote an op-ed piece for the Boston *Globe* about looking for fossils.

> Even today, I can smell the trilobites hiding in the walls
> of gullies carved by the runoff of ten thousand floods.
> As far as I know there isn't a fossil bed within a hun-
> dred miles of where I live now, but the smell of damp
> stones and rotting leaves takes me back to the ravines
> near the town where I grew up. To the hours I spent
> grubbing around in slippery stream beds and climbing

layered walls of rock which cut through the bottoms of ancient seas. My companions were chipmunks, snakes, earthworms and crayfish, but I was looking for older creatures who crept round the edges of my imagination and told me what the Earth was like before human eyes could see.

Digging for fossils wasn't easy work. My fingers stiffen at the thought. I had no tools, only sticks and rocks gathered on the spot. And mostly my fingers, prodding, prying, pulling, trying to free some fragment from the wall and then to crack it open along a natural seam. Trilobites were not common. Mostly I found shells and stalky plant-like creatures. But five or six times I hit upon one looking like a giant bug whose furrowed husk had been preserved for my eyes alone over a quarter-billion years.

On rainy days I get a smell that brings back the excitement of such a find. And what it gives me is a sense of life. The whole organic flow of living things creeping, crawling, swimming across the thin surface of planet Earth, achieving some precarious balance for a time, then giving way. That damp and fusty smell mingles it all together—oak leaves, worms, insects, snakes, chipmunks, fossils, and the boy in the gully—to constitute a show more stirring even than the antics of Popeye or Wile E. Coyote.

That smell defines the space in which I live, uniting me with days and seasons, with rivers and trees, with the living and the dead. One trouble with television is that it has no smell. It is sterile and deodorized. But life gives off a smell. The smell of trilobites binds me to that life and to that encompassing universe within which life fulfills its possibility. That smell rises from the oceans of the past and penetrates into the darkness of the future.

In the fall I sense I belong to all this. I have a time and place in a larger frame. I am a creature of this Earth, a wonder among wonders. Without that sense, what meanings must be borrowed or invented to cover up the emptiness? What cartoons drawn to present the illusion of eternal youth and undying laughter?

For me, Saturday mornings in the fall are a time to encounter the sense of life that gets edited out of school books and TV shows. The sense that gives life meaning because it has to end. It's there in every fallen leaf, in every curling milkweed pod. I know that accepting my own mortality has something to do with what I found when I got out of the house as a kid and opened myself to the world. It has something to do with the smell of trilobites (Saturdays were fun days then, Boston *Globe*, Dec. 11, 1980).

4-8. Killer. At Fort Ord in 1955, the Army did its best to make me a killer. First the cadre of noncommissioned officers (NCOs) who led our company through basic training did their best to repair our sorry physical condition and lack of discipline. Then to make us all marksmen, initially with stationary targets, then using pop-up targets on either side of a squad patrolling along a trail through thick scrub under a hot sun. Eventually we faced the infiltration course, cradling our rifles in our arms, crawling beneath machine-gun fire, first in daylight, then at night; first with dummy ammunition, then with live rounds whizzing overhead. I remember the feel of my elbows scraping against ribbed sand turned to stone. In the end, we attached bayonets and slogged, exhausted, single-file up a hill with a gallows-like frame at the top, a uniformed straw dummy dangling from the frame, an NCO to the side of the frame shouting "Kill, kill, kill!" We were meant to thrust our bayonets lustily into the dummy while yelling "Kill, Kill!" to work ourselves up, but all I could manage was a faint *touché* with the point, and a choked "kill" as an afterthought. It's a good thing I was accepted at Signal Photo School; I am not a killer made or born.

4-9. Shark repellent. While helping dismantle Sarasota Army Airbase in 1946, I got my hands on packets of orange dye used by fliers to heighten their visibility if they had to ditch at sea. Someone told me the dye was also a shark repellent. The question was, what could a thirteen-year-old boy do with several packets of perfectly good orange dye? It struck me that people would not like it in their swimming pools. Thinking of swimming pools, the closest one at Lido Beach Hotel just up the street came to mind. I didn't bother about how the hotel people would feel, I felt it was a great idea. So I scouted the pool, and found I was kept off by a chain-link fence, and would have to lob the dye over the fence like I'd seen hand grenades thrown in the movies. Leaving me, no doubt, with orange hands, which I could shove in my pockets. I didn't want to be tagged as a loiterer, so set D-day for the day after, and went home in gleeful anticipation.

After school the next day, I sauntered aimlessly to the hotel, where guests (Yes!) were splashing in the pool. I looked around for hotel employees. The lifeguard was talking to someone and not watching. I saw my chance, and I took it. I slid the yellow packet out of my pocket, yanked the flap, let dark orange clumps drop into my hand—which magically turned bright orange. I sidled up to the fence, stood sideways, and lobbed the mass toward the pool. Not only did I hit the pool, but its blue waters began to turn bright orange—as orange, say, as my sweaty hands. It was a moment of supreme joy. I walked around to the beach-side of the hotel, lingering for as long as I could stand it (about three minutes) until drawn back to the scene of the deed. The yellow packet was in my pocket, so I kept my hands clasped behind me as if I were accustomed to taking meditative strolls around the hotel. I came face-to-face with a man with rolled-up sleeves who said, "Show me your hands." Saying nothing, I put on a face of offended innocence, and held my hands out of sight. "Your hands," the man said. "Me?" I said. "You!" he said. I never thought of running; the jig was up. Secretly proud, I showed him my hands.

4-10. Making myself happen. The nine examples I have given above illustrate scenes from my life activities as viewed from my point of view. They don't present what happened so much as hint at how I personally scripted what happened during a particular instant or period of engagement. Typically, I depict myself as working alone. These events are largely *my* doing. Only late in life have I realized that I often get confused in the presence of others whose styles of engagement differ from mine, so prefer to go solo to be sure of who I am. That is, to remain recognizably familiar to myself. In a group, I often remain silent as I try to keep up with how others are overtly presenting themselves. I now see clearly that my preferred method of operation is to be fully myself in situations where one person speaks or acts at a time. It is no fluke that I have worked at jobs giving me a high degree of personal autonomy. As teacher, photographer, writer, presenter—I engage without being distracted or interrupted by competing noises and events.

I am predominately a visual person vulnerable to auditory competition. Any sounds in my vicinity (boat horns, calling birds, radios, conversations) distract me from what I am trying to do. Making me extremely inner-directed by choice. I literally lose my mind in cocktail parties, crowds, and large groups. That is, spend all my mental energy trying to suppress incoherent noises that make hash of my thoughts. My lifelong pattern is to seek out gatherings where people take turns, as in classrooms, lecture halls, one-on-one conversations, Quaker worship, libraries, hiking solo or at the tail end of a group where my thoughts and observations are my own.

For me, my chosen profession is to work intently by myself in order to concentrate on what I am trying to do without distraction or interruption. If I can achieve that privileged state, I can be extremely productive. As a side effect, I am often original in being dependent solely on my personal resources, which is my preferred state of being. I am still that little kid refusing to be fed, crying "I can do it all by myself!," now a grownup responsible for his own actions. I love to make little booklets from one sheet of paper by laying out sixteen pages, eight on each side, which—folded, trimmed, stapled—measure 2.75" x 4.25". I write haiku and haiga (haiku with an image), concrete poetry (a hold-over from the 1960s and 1970s), issuing them at Christmas collected into series of handmade booklets. I cook by and for myself, and similarly hike, write, think, and work, making myself happen as I prefer and am able to do. I can be a team player if my share of the work allows me to be on my own. That way I can play a role within a larger group by contributing in complementary fashion without getting distracted or conflicted. If I can't play as myself so I know who I am, I tend to get overwhelmed.

4-11. Action as the meaning of life. My actions are my responses to the questions raised by my situated self in seeking appropriate ways to understand what's happening in the world. So do I engage energy patterns impinging on my senses, ever trying to understand what those arrays can tell me about my current situation. Round and round go those energies, engagement after engagement, prompting me to adapt to my surroundings by acting in meaningful ways in response to the sensory patterns I perceive.

If separated for only a few moments from the biological environment that supports me, I cannot maintain myself, so am committed from birth to a lifelong course of ecological engagement. A commitment on two levels at once, unconscious and conscious. The *unconscious* level handles energy states my body can adjust to on its own, fine-tuning levels of carbon dioxide in my blood, blood pressure, temperature, heart rate, digestion, hormone secretion, and many other activities crucial to my current well-being. The *conscious* level is required for states that challenge me to figure out what to do next because I've seldom or never been in that position before. Action is the payoff of my sensory engagement with life, whether I arrive at it by one route or the other. Everything hangs on the actions I venture, so I stand alert to gauge whether they are successful or not.

4—ACTION

Action, I propose, is the meaning of our individual lives, action *appropriate* to particular life situations, *effective* in meeting the degree of challenge we face, and, I would say, *caring* for all participants in a given engagement. Perception gauges where we stand; situated consciousness fits that standing into our spectrum of prior experience; action itself advances that standing in either an automatic (unconscious) or a considered (conscious) manner.

Beyond the level of individual acts, actions dictated by cultural traditions, beliefs, and procedures set the collective meanings of our individual actions taken together in responding to issues of common concern—the Japanese attack on Pearl Harbor, felling of the Twin Towers, periodic stock market crashes, droughts, pandemics, deaths of national leaders, or governments at war with their people as lately in Libya and Egypt, currently in Syria. The challenge to a culture is to provide a social structure adequate to the occasions on which group action is advisable or necessary. Families, tribes, cities, and nations operate on their respective levels of awareness, so attempting to assure the well-being of their members as a group.

Ideally, a top-down social structure requires those at the pinnacle to be sensitive to the actual needs of those at the base of society. A grass-roots social structure builds outward and upward from the situated individual at the base to those who represent and address the collective needs of all members of a given social structure. By my view, the best society is one in which each works for the benefit of all, and all work for each individual. When an elite group commandeers a nation, coopting its wealth, actions, and benefits, the people suffer deprivation, which is made worse if they do not engage in their own defense. The most beneficial structure is supported by, and provides for, all people to equal degree. What else is a society for if not a hundred percent of its people? Again, true in the ideal, but in practice something else again.

4-12. Judgment. Welling up from our interiors, judgments reflect our subjective verdicts on the impressions we form from the energies affecting our sensory receptors. Such judgments reveal the valence of our feelings toward those impressions—whether we deem them good or bad, desirable or undesirable, noble or evil, welcome or feared—from our situated points of view. Those valences, in turn, predetermine how we will react on a given occasion—pro or con, yea or nay, hug or shove, smile or frown. Indifference and ambivalence are mental states between two opposing extremes in which we could go either (or neither) way. In that state we withhold judgment until able to form a clearer picture of what's happening.

In the realm of politics, it is no accident that votes for easily distinguished rival candidates are often so evenly matched, and truly undecided voters are so few: since our survival depends on taking definitive action, our brains go to great lengths to paint any situation in black or white, sharpening distinctions, making choices obvious—lumping actual differences to make the decision all the easier, distorting or overlooking subtleties for the sake of maintaining a stark difference between candidates who skillfully dress their remarks to appeal to what they see as their key—and often incompatible—constituencies.

Judgment by the situated self is positioned at the mental junction where attitudes toward sensory impressions are clarified in preparation for subsequent action, reducing the question to a simple choice between should I or shouldn't I? Once resolved either way, the rest is planning and action—or perhaps resolute inaction.

4-13. Decisions. There being more than one way to bring about a desired result, planning begins with a review of possible options, and a preference for which one to act on. Lists of pros and cons are drawn up, weighted, evaluated, and choices decided. Judgment clarifies the overall appeal or attractiveness of a given impression within a specific situation, then decision initiates planning by selecting a particular course of action from among multiple alternatives. In my case, my self-deliberations often go on for months or years until I come to clarity on how I want to proceed. I worry decisions in my head until harmony descends upon me as if from above, nagging doubts subside, and I resolve to act. When the time is right, I make my move—putting in an outhaul for my boat, a new roof on my cabin, my thoughts in workable succession, my house in order.

4-14. Goals. In deciding on the spot what to have for breakfast or what movie to see, prior planning promotes action as guided by goals set beforehand. Those goals reflect judgments and decisions made by situated selves reflecting their degree of approval or disapproval of a meaningful set of sensory impressions interacting with other impressions as recalled from memory. Considered action, then, is guided by felt judgments upon sensory impressions recognized as being both meaningful and desirable. Instinctive or spontaneous action follows a more direct route paved by set habits and routines, bypassing conscious consideration and judgment altogether. In *Consciousness: The Book,* I wrote:

> Goals direct attention to what needs tending to as a means of channeling our vital energies toward a version of the future we prefer over other alternatives. In setting goals for ourselves, we anticipate what might happen, and put our energies into realizing the possibility that appeals to us most, or considering our means, is at least within reach (page 186).

> Life, then, is a projective personality test. The work never ends, so the goals we set ourselves (as expressions of our values) tell who we are at that time. Which kindles the remarkable side effect of instilling hope by working not just for today but for tomorrow and the day after. Setting goals is one of our ways of working to build a future to our liking. . . . Goals, that is, are a means of turning our lives into self-fulfilling prophecies. We venture the goal and fulfill ourselves in striv-

ing to achieve it. We may not end up where we thought we would, but the adventure has been ours the whole way (page 189).

The imagination we rely on in planning the future takes memory to store our intentions, otherwise we would surely forget them as more pressing issues come up. Judgments, decisions, goals—all depend on memory to hold in mind what we plan to do. At a minimum—as in fast action—we need to stretch memory out for several seconds—enough to keep repeating a phone number so we can put down the directory and dial the call. Goals are placeholders in memory, which are useful in walking along a trail so we can remember which way we were heading when we stopped to tie our shoe. Without goals, we'd be trapped in the now, the now, forever in the now. We would be spared a great deal of worry, perhaps, but at the cost of an inability to work toward a definite future.

Our loops of engagement are a function of memory in allowing us to shift our attention from sensory phenomena to figuring out what they mean, planning accordingly, acting, and then revising our impressions in imagining what next steps to take. I view so-called working memory as the minimum span of nerve potentiation allowing us to keep in mind that last thought of what we were going to do, enabling us to see how we did, and plan and conduct a next round of revised action. This, I believe, is the source of our respective streams of consciousness in driving our shifting attention from one moment to the next. Long-term memory requires more than momentary retention—long enough to build lasting connections between neurons as urged on by diligent repetition (such as flipping through flash cards, for example, or repeating verses of songs at frequent intervals) or enduring emotional stimulation (episodes of fright, joy, dread, desire, and so on). I'm still working to understand my own mind, a job I undertook in 1978. That suggests the mental power by which goals can be set and acted upon.

4-15. Projects. We sort the several tasks we hope to accomplish on a given day into separate bins in our minds so we can focus on each of them one-at-a-time in order to have progress toward a number of goals to show for our efforts at the end of the day. If we don't apportion our attention in systematic fashion, we are apt to waste time wondering what to do next, where to do it, how to do it, who to call for help, where to get supplies—running the risk of making no progress toward any one of the goals we set ourselves, but fail to pursue in a direct manner. Projects are mental workshops where we go to focus our time and effort in putting our limited resources to good purpose. Knitting a scarf is a unitary project, as is building a ship model, drawing up a contract, throwing a party, having a baby, building a house, deciding where to retire.

I try to schedule work requiring originality and creative thinking for early in the day when my wits are freshest, holding lesser tasks in reserve for later in the day or evening when I may not be at my best and a routine performance will do. I frame each day with automatic routines—showering, getting dressed, eating breakfast. Then I move briskly into writing, working on a PowerPoint, scheduling appointments for later in the week, eventually answering emails, shoveling the walk, cooking lunch, reading an article, going shopping, calling the garage—and the day is gone, leaving dangling threads to be picked up the next day.

When working on a major project with a deadline, I ignore every other claim on my attention by shifting into get-it-done, nothing-else-matters gear while neglecting every other claim on my attention. I concentrate best in single-minded mode by declaring one priority absolute, fully intending to get to everything else when I'm done. In the meantime, another priority creeps up behind the first one, and jumps in to take over my attention, leaving lesser matters to fend for themselves, often becoming moot because it's too late to do anything about them. Which is why every surface in my apartment is cluttered with piles of paper waiting to be sorted because the sorter-in-chief always finds better things to do than get organized.

In my case, multitasking is a myth. I stack parallel tasks I could be working on in serial order, tending (or not) to each in its own time. Typically, the brain is visualized as a network of parallel modules all working away at once. Here I place more emphasis on serial loops of attention that follow one another in sequence, each module (I call those I am aware of *elements* or *dimensions* of consciousness) having its own place and playing a different role in the queue. As I see it, *attention* is the master of ceremonies in my mind, and its dictum is that three-ring circuses are invitations to distraction, so my stream of consciousness is governed by a single stream of attention, not an array of parallel streams. My unconscious mind may operate in the background on several fronts at once, but if I elect to focus on one front in particular, my conscious mind must choose between them so that each has my exclusive attention for itself. I suspect people vary widely in this regard, some being able to split their attention, others (like me) being helplessly single-minded.

So what is this driving force of attention that compartmentalizes my mind into different projects or areas of concern, and restricts me to dealing with them single file? As I view it, attention is the narrow turnstile (or neck of the hourglass) that allows entrance into awareness on the basis of urgency, salience, gravity, or importance within the frame of a particular situation. And that urgency is determined by comparison so that only those candidates exceeding or falling short of what might be expected in that situation are allowed in *because they have behavioral implications.* That is, they have weight in regard to my personal hopes, dreams, or assurance of attaining a particular goal. I sort such candidates for attention by the degree to which they affirm or dash my hopes, conveying the message that I am demonstrably either on course for success or headed in the wrong direction and am in danger of ending up on the rocks. The candidate that gets my notice is the one I least expected, so must be dealt with right away. Since the goals I set have survival value — giving up smoking, going on a diet, falling in love, getting a job—I monitor my approach to them with great interest. That is, I compare my current position against my destination in terms of goals, wishes, or expectancies, and admit those notably above or below that standard for further scrutiny and refinement. As I put it in *Consciousness: The Book:*

> In using projects to address the outside world, we trans-late our inner experiences into outward acts intended to achieve a certain effect. That effect is then translated back into the language of our nervous system by our senses as guided by our expectancy at the moment, to be processed in a neural network that sharpens its fea-tures so we are able to compare patterns we hoped to bring about against the patterns which actually appear. Comparison is at the heart not only of consciousness but of the loop of engagement connecting our inner and outer worlds in terms of the situation we picture our-selves being engaged in then and there. If both the ex-pected and actual version fall within a range of congru-ity we can accept, we move on to the next stage of the project; if not, we refine our attention and motor con-trol, and try again (p. 205).

Projects, then, are like distinct stages or legs of a much longer journey. We want to reach Boston by nightfall, not go all the way to Washington; or we're working on a chapter, not the great American novel. So do we face into eternity, one goal, one project, one action at a time.

4-16. Relationships. Some people work on projects while others build relationships. As I see them, relationships are a kind of project in the social world. Or conversely, projects are a kind of relationship in the physical world of matter and energy. In my personal experience, I find women commonly building relation-ships (being with someone) while men are more likely to work on projects (playing poker, mowing the lawn). Both are goal-oriented activities, one leading to families, schools, friendships, care centers, and communities, the other to roads, bridges, cars, lawsuits, and corporations. Both build larger wholes from smaller units or parts, so it strikes me that the mental skills and attitudes involved are largely similar, it's the media in which those skills are expressed that differ.

O.K., I've dug my grave, now I must lie in it. Of course men as well as women can and do work in both social and physical media. But my general impression is that women are more likely to go into teaching, nursing, and public assisting, men to go into construction and manufacturing. Women can give birth while men can't. Women feed babies at their breasts, sing lullabies to them, teach them to smile, to talk, and to cook. What are men good for? Bringing home the bacon, building houses, driving cars and trucks, playing games, investing, fighting wars. Both are equally responsible for their respective records of achievement. Social relationships and active projects are equally necessary to survival at individual and communal levels. Both men and women form attachments to their families as well as to the outcomes of their own projects. Painters, musicians, singers, and dancers of either sex are sensitive to both social and physical relationships.

Even so, speaking generally, I often observe a difference in temperament between men and women, which suits them to different social roles and ways of life. Many women who base their professional careers on projects come around to having children later on, and many men turn out to be excellent caregivers. My bet is that hormones are responsible for the two temperaments I observe. Both men and women are exposed to hormones in the womb, and have varying levels of estrogen as well as testosterone in their bodies. Being stuck with the body I have, I can only report what I observe in the actions and deeds of other people of all sexes (there are way more than two alternatives).

When my mother and father got married, my father said there was room for only one career in the family, so my mother, well on her way to earning a doctorate in geology, had to focus on raising children and becoming a Sunday painter in her spare time. That, I believe, was a matter of cultural tradition, not hormones. The home sphere and work sphere were rigidly separated. The domestic playing field is more level now than it used to be. Hampered by having to look through a man's eyes, I want to be fair to women who look through eyes of their own. So I emphasize the role of building relationships as playing a role in human behavior as an alternative to projects by way of acknowledging the thinness of my personal experience.

4-17. Tools and accessories. Tools largely determine what we can accomplish in this world, and tools, like accessories, are the gift of the cultures we are born to, so have a huge impact on the people we become, the work we do, the skills we develop, and our actions in this life. To become an astronaut, a nurse, a fireman, or a chef, we have to develop mastery of many tools that define who we are. Think how different it would have been to live in the stone or bronze age, how that age would have limited our possibilities through the assortment of tools available to us at that time.

George Eastman invented photographic film in 1885. Which led to motion picture film, which led to roll film for still cameras, which is where I entered the picture. My first camera was a little Bakelite job with one roll of film I got from a cereal company for twenty-five cents and a box top. That purchase eventually led me into the darkroom with its typical array of tools and chemicals. That entire era is now history, those tools are fast becoming forgotten artifacts from an earlier day, and I have gone digital. I was born in the year George Eastman died. He made millions in his time, but now even his once-lucrative inventions are fading away.

When I was a kid, crews of men used shovels to dig ditches and graves; now those jobs are done by backhoes with one man at the controls. Except for farming, muscle-power has become obsolete. We commonly use heavy machinery and high-tech gadgets to make ourselves happen these days. Smartphones, computers, tablets, iPods, GPS units, hybrid cars, calculators, satellites, particle accelerators—people are doing more in a year by tapping their fingertips than has been done by all-out bodily labor, skilled or otherwise, in all the ages of history before now. Erect a pyramid? Give me a few days and we'll build it on-site to your specs. A skyscraper? Say, in sixty days. Locate livable planets and likely colonies of extraterrestrial life? We'll have the maps in a year or two.

In 490 B.C., Pheidippides died after running the twenty-two miles from Marathon to Athens with news that the Greeks had beaten back the Persian invaders; now he could post "Rejoice, we conquer!" on Twitter or Facebook and the whole world would know in a few seconds. If our tools change, we change, for how we make ourselves happen in the world is a big part of our identities. At the current rate of technological development, the tools we master become obsolete in a matter of months. If we don't upgrade ourselves to match their replacements, we find ourselves becoming obsolete as well. If we can't keep up, we're not in the running.

Some tools and accessories distort our engagements, rendering them crude or ineffective. Firearms give each of us the power to kill people we don't get along with, precluding the possibility of engagement. So-called recreational drugs achieve the same end by altering the chemistry of engagement itself. Alcohol and tobacco may seemingly ease social tensions while posing other dangers to our physical well-being and perhaps that of friends and family in ways we cannot imagine. We resort to elective cosmetic surgery to make us more acceptable to ourselves and others, but putting on a new face or figure does not hide or override our genes any more than our clothing does. In that sense, we engage on the basis of false pretenses, which puts more than our vanity at risk. Seeing countless tourists walk the streets of Bar Harbor with cellphones clamped to their ears makes me wonder about the wisdom of disconnecting our attention from our physical bodies. Yes, it allows quasi engagement at a distance, but prevents us from being fully where we claim to be. Why come to Bar Harbor if the emotional high points are spent on the phone? As I say, I have difficulty with multitasking, preferring a wholehearted commitment to keeping my presence, attention, and values intact.

4-18. Skills. As an increasingly greater part of our activities must be spent just keeping up with the times rather than doing productive work, we need to go back to school more often to upgrade our skills. The world we are born to has always become a new world every morning, but the rate of change is picking up so rapidly, the world has to spin faster on its axis to keep up. Or we need more days in a week. Otherwise we need to keep retraining ourselves effectively at a faster and faster pace just to stay in the game. I see this as largely the result of growing competition for the limited number of niches the electronic-manufacturing-service-military-business economy is able to support. Companies really hustle to grab a piece of the pie. Too, as global population grows, new niches open where least expected, adding to the worldwide frenzy of innovation.

The question is, will Earth be able to keep up with the demand for new goods and the corollary skills they require of us? What's good for R&D is not necessarily good for either us or our planet. Once, Earth was in control of us, and we had no option but to adapt to its style in order to survive. Now, we act as if we were in control, expecting Earth to humor our every whim and desire, whether or not it has the resources to spare, or can tolerate the resulting depletion, pollution, and degradation of our common living space. Fracking is a contemporary case in point, giant corporations laying the infrastructure as if it were the proven way of the future. Inadvertently, our human life skills are becoming death skills, with the outcome being our extreme success in making our planet unlivable. Humanity, truly thy name is Entropy for wasting that which is most precious to you. As we live by our skills, so do we die by those very same skills.

The point being that having lived for so long on Earth's bounty, when that bounty runs out, we have no place to hide. Our skills will be worthless because they can no longer sustain us. Earth's love for each of its children will be fully and irreversibly expended. What, then, are we to do with ourselves, since our skillful actions are the true end and essence of our being? That, now, is the question we face as we near the end of the road. For, demonstrably, Earth has neither the patience nor the ability to keep up with our demands and expectations.

When the grid goes down and the Internet crashes, will we remember how to slow down enough to pick up the old tools and use our muscles to do the work that needs to be done? Perhaps we will recover the art of making stone tools, and the art of using fire to heat our food and warm our bodies. We can always raid the ubiquitous trash dumps we have planted across every landscape, but that will undoubtedly prove to be hazardous to our health. Live by our wits then? Live off the land? Live by our skills? Live off the sheer will to survive? Not very likely. Unless . . . .

Unless we can develop the entirely new skill set of *getting to know ourselves—our conscious minds* and what we are doing before we reach that insurmountable barrier up ahead marking the end of the human adventure. There is a way to avoid catastrophe. One way. The inner way toward knowing ourselves and understanding why we make ourselves happen in the world as we do. Once we get there, and each does her bit, perhaps together we can cultivate the will to save the world for all of nature's children. ◯

Chapter Five

THE SITUATED SELF

5-1. Outlook. Sensory patterns don't just come to us through our senses, we shape them to fit our fears, needs, and desires so we see what is important to us. That is, we are motivated to notice certain sensory details available to us and not others because we have found such details affecting in the past. That is what I mean by having an *outlook*—a certain take on the world because we are who we are by living out our past histories of experience.

Unconsciously (yet often creatively) we form patterns from a host of seemingly unrelated perceptual features given coherence by the detailed geometry of our sensory receptors, then face the next challenge of determining what those hard-won patterns might mean in terms of the situation we are engaged in at the moment. So what? we ask ourselves, what difference does it make?

Our sense of being in a situation is based on the positive and negative feelings that various patterns stir up in us, together with the biological values we put into play, and the role of memory in recognizing (reaching out to and finding familiar) the patterns we are concerned with—all adding to a felt understanding of the situation we are facing as based on our sensory impressions at the time.

Sensory patterns demand to be interpreted as examples of this or that type of experience. We don't experience the shape or nature of a duck, say, so much as the concrete duckness of a duck. If it looks like a duck, walks like a duck, quacks like a duck, surely it is a duck and nothing that merely resembles a duck (unless it is a decoy, model, image, or representation, in which case it is something other than what it might seem at first glance).

Outlooks and situations resolve all such considerations by combining them into an understanding of *what-is-ness* and *how-ness* and *why-ness* that includes both the pattern and the perceiver so that their relationship is patently clear and understood as meaningful to the one particular person who is moved to pay attention for reasons of her own because she is who she is.

The pattern emerges within a situation as not just a collection of individual features and details so much as the inherent relationship between those details (including the beholder's personal history) so that the situation is grasped and understood as a dynamic engagement pointing the way to appropriate action. If it's a duck, shoot it; throw it a crust of bread; make sympathetic quacking sounds; point your finger and say, "Oh, look, there's a duck."

Situations as we perceive them are at the core of our looping engagements with the world. Given our individual outlooks, they are the best we can do in figuring out what sort of world we are facing right now. We receive energy from that world, true, but not the world as it is in itself. Based on our habits, expectations, opinions, and experience, we piece together one version of all possible worlds, and for the moment, that is the operative situation within which we act as we do.

Situations are what we are able to make of sensory impressions as fleshed-out with our feelings, values, memories, and understanding of what sort of a scene we are likely to be facing. Always, always, always, situations reflect our opinion of what we have gotten ourselves into because they are centered on our personal experience and the outlook we have earned through enduring a lifetime of joys and hard knocks.

Regarding the loops of personal engagement we experience in living our lives, our first task is to form a sensory impression of the world based on our expectations, arousal, interests, attention, and need for clarity at one level of sensory detail or another. Then a second task is to combine our feelings, values, and opinions into a situation that would make the sensory pattern we come up with meaningful in light of the lives we have lived up till now. Setting up a third stage of our looping engagement in which to act appropriately within the situation we have constructed for ourselves out of bits and pieces of the past lives that have gotten us this far.

Round and round we go, engaging first one situation then another, always striving, always learning, always trusting memory, opinion, and imagination to show us the way. Forming clear sensory patterns, putting them in the context of plausible situations, then acting as we are moved and have the opportunity to do, so advancing to the next round of our engagement, and hopefully the round after that, making ourselves happen in the world in response to the flow of energies impinging upon us from that world. (Based on "Reflection 292: Outlook," posted to my blog, *Consciousness: The Inside Story*, July 13, 2012.)

5-2. Inlook. In the 1939 movie *The Wizard of Oz*, it was Toto the dog who revealed the Great Oz to be a humbug. Dorothy had finally gotten an audience with the public persona of the wizard, which was all flames, smoke, and amplification, when little Toto pulled back the curtain, revealing the wizard to be a mere mortal at his console behind the scene. "Pay no attention to that man behind the curtain," said the man behind the curtain into his microphone, desperately trying to shift Dorothy's attention back to his special effects. So did Dorothy's situation as a supplicant before the Great Oz abruptly morph into a different situation altogether. Much as Lance Armstrong's situation as seven-times winner of the *Tour de France* morphed into a sham when he told Oprah Winfrey and her audience of millions that his doping scheme was an effort to "level the playing field" so he'd have a fair chance (because everybody else used dope, so why shouldn't he?). Often, it turns out, situations are not what they seem, there being one for public consumption, and another for private practice.

Think of President Clinton declaring, "I never had sex with that woman." Think of Newt Gingrich having an extramarital affair as he crucified Clinton for the same offense. Think of Bernard Madoff (Bernie to his friends) bilking his clients out of their savings, President Reagan denying knowledge of the Iran-

Contra affair, President Nixon excusing the Watergate break-in, and child molesters everywhere worming their way into parents' trust to gain intimate access to their children, as a former doctor at Children's Hospital in Boston used his professional standing to examine hundreds of children in private, including mine.

Situations are not always what they seem, which is why I take pains to say that they are *construed* or *constructed* for the sake of personal comfort and consumption. Let the constructor be wary of duping himself as well as his public by looking toward the motives and values within which his actions are situated — *truly* situated.

5-3. Natural cycles. In 2006, I attended a month-long research seminar run by the Quaker Institute for the Future. My project aimed at trying to get fishermen and fisheries managers to at least begin their discussions on the same page. I had been to countless management meetings in which managers (generally having biological backgrounds) were focused on the impacts of fisheries themselves on the stocks that were caught, while fishermen saw everything in terms of natural cycles that were responsible for the ups and downs in the number of fish that they landed. The arguments went round and round, never making headway toward more effective fisheries management. In the meantime, populations of everything but lobsters went down, and down, and down all along the Maine coast.

First I made an ambitious PowerPoint presentation about stewardship by all parties being the only way to bring back populations of cod, haddock, hake, redfish, flounder, etc., from being heavily overfished in the last century. But I found that one thing fishermen stay away from is meetings based on Power-Points, so I switched to writing a book about consciousness, trying to nail down just why different people come at issues from such entrenched positions. I now conclude that we speak from situations as we construe them for ourselves, not from facts. The conflict lies in our inherent intransigence, our inability to question positions based on a lifetime of personal experience.

Fishermen and fisheries biologists have earned their credentials by the widely divergent routes that have made them who they are. There's no going back, no chance to revise history. We are what our experience has made of us, period. The situation we are in is an emblem of who we are. Questioning our situation comes across as an attack on ourselves.

The same is true on both sides of the argument about global warming. Skepticism is built into the life histories of those who don't believe in it, while the urgent need to act as quickly as possible makes sense to those familiar with melting glaciers, receding sea ice, increasing turbulence, longer growing seasons, and species migrations, among many other indicators. Where believers see the buildup of greenhouse gases in the atmosphere as a smoking gun, their opponents discover natural cycles that have happened over and over again, so there's no need to worry. Again, the difference points to how we build situations from "facts" we are personally comfortable with.

5-4. Lincoln Park. Ten-thirty, time to go home. I am in the darkroom, as I often am these days, making prints, waiting for the last ones to come off the dryer. It's been a long day. I put the prints in a yellow Kodak box, gather my gear, turn off the lights, shut the door. I love the three-mile nighttime walk from Garden Street, through Harvard Square, down Mass. and Putnam Avenues to Magazine and finally Tufts Street in Cambridgeport toward the Polaroid Factory. I unwind as I walk through misty shadows. Lincoln Park is the still center of a cyclone of traffic speeding around the periphery. A few couples are strolling along crisscrossing paths. When I come abreast of Abe on his plinth in the dark heart of the park, I notice a couple strolling toward me. I don't want to appear menacing toward them, so start humming softly. As we near one another, I see it's two guys. The guy on the right starts running as if to pass, so I move left to make room. As he comes up to me, he grabs me around the neck and swings behind me, pulling me against his chest, arms locked at my sides. The other guy comes up and starts bashing me in the face. Surprising myself, I hurl curses I learned in the Army at the two faceless men — *mother*-this and *chicken*-that. The one keeps bashing, I keep swearing. Other strollers keep strolling. Then the pair runs off. Stunned, I pick prints off the path where they spilled from the box when I dropped it, and continue home, stopping at Cambridge Police Station to report what had happened, to be told there was nothing they could do. For a month, proven dupe that I am, I clutch a monkey-wrench out of sight up my left sleeve, then I figure I was just the wrong guy in the wrong place at the wrong time, and let the wrench lie in my toolbox.

5-5. Looking for Moses Butler. I know right where he's buried. I came across his grave in about 1994 while walking from where I hauled my boat up on Butler Point back to my car by the camp on the south end. I'd been told that you used to be able to see his grave from upstairs in the house (since removed) next to Hope Butler's, before oaks had grown into the former blueberry land, blocking the view to the graves by the big rock. I had in mind the pattern the three graves made — two next to each other, one above the right one (looking north). Moses Butler, I'd been told, next to his Indian wife, with her father in line with her.

Anyway, when Emery calls for the historical society in 2009, I tell him I'd be glad to show them the site. So I drive over from Bar Harbor, meet the three folks from the society by the gate on the road to the point, and drive down the ruts past the edge of the field to the camp. From there we walk through a ferny wet area to the base of a rise, and at the top by a medium-sized spruce, I sweep my hand and say, "It's right around here." Well, the oaks had filled in the barren, dropping leaves every year for almost fifteen years, burying the graves — really just rows of smooth, flat boulders lined up in the sod, all three built the same. Not lined up anywhere that we can see now because they're under the leaves. For two hours, the four of us do our best to scrape away leaves at the spruce end of the rise using the sides of our feet, without luck. The big rock is still there, on the far end. I'm sure the graves are somewhere between the spruce and that rock, probably closer to here than to there. We set a date to try again with rakes and shovels, and improvised metal prods to poke into the soil if we need them.

Four or five of us show up a few days later, use string to mark out areas to rake, and do the best we can to get down to the surface. No graves anywhere in sight. So we poke with sticks, rebar, and pointed tools. I keep saying, "They've got to be here," but if they are, in two hours we don't find them. I come back a third day and rake toward the big rock, with no luck. Emery and I come back a year later, ditto. I come back one more time. After that, we lose interest, or more honestly, the will to rake, poke, and dig. We all want to find those graves because Moses Butler was—after the Indians and the French—the first English settler of Franklin, Maine. Kenneth Roberts had written up the adventures of Moses Senior in sailing to the New World via Boon Island off Kittery, where the ship was wrecked, crew and passengers almost starving or freezing to death in sight of the coast. Moses Junior was a local celebrity, who'd threatened to move on when, looking out his window, he saw a light from another encampment across the bay—too close for his comfort. But he'd stuck it out, and was buried on this land.

The upshot of our amateur archeological dig being that the graves are not just misplaced but confirmed lost, and might be forever. The prospect of finding them has gone in stages from good to bad, dragging my confidence with them. I directed the search, and had come up empty-handed. In desperation, I ask the owner of an infrared heat detector to help scan the ground, on the theory that the rocks just below the surface would show up as a cooler area, and the pattern would reveal the graves. But reflection from leaves at ground level produces so much confusion, there are no patterns to be seen, just noise. I lose not only confidence but face among local historians. I don't have the heart to nobly keep searching. I find it a very disheartening situation to be in.

5-6. Will somebody tie my shoes? I remember sitting on the porch in the spring, having my shoes tied. That would be in the Maple Avenue house, probably in 1934 or 1935. It is warm and sunny, I remember that. I even remember the bench we are sitting on. But the difference this time is I am sitting in my father's lap. *He* is tying my shoes, not my mother. Not only that, but *I* am sitting in his lap—which is what makes the occasion so memorable. I note it at the time, and will never forget the scene. It is a unique situation in my young experience. In our house (as I learned from my mother much later on), due to the teachings of John B. Watson, founder of behaviorism, laps are not for children to sit in (*see* Reflections 3-18 *above*, 6-11 *below*).

5-7. Helping hand. Equally clear is another scene from about the same time, in the living room, when my mother slips and falls on a newly waxed floor. Hearing the sound, my father comes out of his study and offers her his hand, being remarkable for his actually touching her in my presence. In my family experience, I'd never seen such a thing.

5-8. Buried remains. Fresh out of high school, I'm in Nespelum, Washington, on the banks of the Columbia River. There are clouds in the sky, but I'm not looking at them, or at the river. I'm too busy digging a hole in the ground. Looking for Indian artifacts on the Nespelum Reservation. The Chief Joseph Dam is under construction, and this ground will be flooded. I'm a volunteer with a team of archaeologists from the University of Washington. Three feet down, I think I've found something. Hard, white. I switch from trowel to whisk broom and toothbrush. Whisk, whisk; brush, brush. There's a suture. Looks like a skull. Brush, brush; blow, blow. See, it's rounded, like a dome. Gotta be a skull. We haven't found any human remains on this dig. I'm gonna be first! Brush, brush. The dome has a funny edge. A ridge, like Neanderthals had. This has gotta be really old. After what felt like hours of brushing away a few grains of sand at a time, I have much of the dome exposed, ridge, eye socket, and all. A real archaeologist comes by to see how I'm doing. "Whatcha-got there, Steve? Looks like some kind of turtle." The "eye" socket is where a leg once poked out. (Based on *Consciousness: The Book,* pages 10-11).

5-9. Uncle Sam's helper. Thirteen and in eighth grade, I live for a year on Lido Key in Sarasota, Florida, and help Uncle Sam shut down the local Army Airbase. I have a good sense of geography, and figure out a route from a bus stop along a seldom-used rail line to the dump in the southeast corner of the base. The wire fence can easily be lifted, giving access to acres of what I think of as Army surplus the government doesn't need now that the war is over. Which is how I also present the situation to my parents. On Saturdays I ride with Russ Shinn, the bus driver, out to the stop by the tracks, and hike along the rail line through a swamp to the fenced corner of the base. I travel lightly, but have screwdriver and pliers in my pocket. For my troubles, I lug home gyroscopes, radios, bucket seats, dummy bombs, smoke grenades, and stuff I spend hours taking apart to find out what's inside.

One time, I pile my loot outside the fence, load myself up, and trudge down the tracks—only to feel, then see, a locomotive coming my way at the far end of the swamp. I understand immediately that the track on its raised bed isn't wide enough for both of us. I don't want to retreat, and the slopes on either hand are so steep I would end up in the water if I stepped aside. My plan is to get to the trestle halfway into the swamp and slip under the tracks before the engine and I meet up. I speed up my pace, metal parts banging against ankles and legs, but it's slow going anyway. The train gets closer, clearer, bigger; I see steam venting from pistons on either side. I don't want to get caught in that steam, so I hobble as fast as I can, clanking the whole way. The engine keeps coming. We reach opposite ends of the trestle at the same time; I get a whiff of oily steam as I slide down the bank with my booty—and it's over. The train passes. I find myself lying on my back, breathing hard. It was a close call, followed by a passionate sense of relief. I don't ease my grip on my treasures, pick myself up, and reach the bus stop just as Russ is pulling in on his hourly rounds, right on schedule.

5-10. Belief. At the heart of each of the six situations I have described in Reflections 5-4 through 5-9 (getting mugged, looking for Moses Butler, sitting in my father's lap, he helping my mother up off the floor, excavating a turtle, and dismantling an airbase), I have a strong belief that I am observing or engaging the world as it really exists at a passionate moment of truth. That's what I refer to as a felt situation, an event in which I wholeheartedly engage as fully myself. Affect plays a prominent role in every felt situation because strong beliefs and feelings are paramount. The word "belief" has overtones of love, caring, and desire for how the world *should be*, not necessarily how it *is*. Belief, whether ironically or not, says more about the believer than the state of the world. I didn't want the pair coming toward me in the dark to think I might be a threat to *them*. I thought I could take members of the historical society straight to the graves. I recognized how unusual it was to sit in my father's lap, shedding light on the usual state of affairs. And how unusual for my parents to touch each other in my view when confronted with the fact. The skull-like quality of the turtle carapace lay in my inexperienced mind. And walking the tracks loaded with booty was more of an adventure than I bargained for. In each case, I exposed the assumed structure of a situation to be other than I passionately believed, leaving me in a state of wonderment at my true situation as revised. Not just standing there, but busy altering the shape of my beliefs as sponsored by the neural network in my brain to make sure I would remember the day's lesson

5-11. Dream situations. My dreams are all about situations in which I am aware, but can neither perceive nor act. Typically, I am trying to go somewhere or do something, but get turned around so my good intentions are scrambled and I end up lost, confused, and full of regret. When I fly off the ground by flapping my dream arms with great effort, I immediately find myself tangled in power lines crisscrossing the sky. Once past the subway turnstile (for which I have the wrong token), I end up on a train heading away from my true destination. Having promised the head of the department to teach a certain class, I find myself showing up for the first time at the end of the term. That is, my dreams reflect the true fact that I can't perceive and can't act, leaving me frustrated at being unable to engage as I want, facing myself as the one responsible for that state of affairs.

Recently I've had two dreams of an entirely different sort. Happy dreams; dreams of competence and success, but equally fanciful. In one, I offer my services as a counselor to the principle of a school and—wonder of wonders—he accepts. I warn him I won't sit in my office behind a desk meeting students by appointment, but will engage them wherever I find them around the school—on the playground, in the gym or cafeteria, on the street corner after school. He says that is fine with him, however I want to do it. So I just walk up to groups of kids and start talking with them, dealing with them directly on their terms, not mine, so that we all engage as fully ourselves and don't play games of hide-and-seek. The kids are happy, the principle is happy, I am happy. I'm not used to happy-happy-happy dreams, so see my dreamself exploring a new direction.

This is backed up by another dream that starts out as one of my usual nighttime bouts of self-blaming as the one responsible for failed hopes and expectations, but then turns out happy, even celebratory. It begins with a tense situation in which my dreamwife is inexplicably getting married to someone else, and I am trying to figure out what that is all about, and why she is doing it. We are at a large, open-air gathering, milling around, which is where many of my dreams seem to end. But this time I put on an eighteenth-century costume with white stockings, heavy black brocade coat with tails and elaborate gold piping, and eventually a powdered wig with a pigtail. I sit at the front of some kind of conveyance full of partygoers, and act as a kind of major-domo or ringmaster, heading-up a parade. I keep holding up strange pieces of old costumes and asking what this might be for. I am jolly with everyone I meet, and am well-received by all, like the counselor in my earlier dream. Here I am accepting and even enjoying myself in my dreams, playing myself to the hilt, being as accepted by others as I am accepting of them.

These dream situations illustrate the passionate element of loving, caring, and desiring being acted out while I am asleep. Acted out by whom? None other than my dreamself, the core self or persona I am whether asleep or awake. More-or-less effectively when awake to the world, fancifully when asleep. I am the one who makes himself happen within the strictures of whatever situations he finds himself in. Particularly situations in which he is emotionally involved and highly invested, as on professional and intimate family occasions.

5-12. Subject and object. The tricky thing about being in a situation is that you are really facing off with yourself because both the inner and outer poles of the situation are your own doing. You are the subject and also the object of any situation because both are firmly seated in your own mind as complementary self and antiself. That is why, at the same time, you can rise and seemingly look down on yourself and the scene you are facing, as is so common in near-death experiences, and is such a common feature of dreams where you are wholly responsible for being yourself and for the shape you give the world around you. What the world does supply is patterns of energy, but what such a sensory array feels like and means is your contribution as based on your lived experience, either in the past or your current imagination as derived from that past.

Just as my dreamself and the scene I am engaged in are both my doing while asleep (even if nearly awake), so my dayself when awake and its seemingly external felt situation are also both mine as the two poles of my subjective awareness, one clearly in my head, but the other also in my head as projected onto patterns of energy streaming from the world around me. When I read a book, for instance, from my point of view I am both the reader and the writer of that book, for I understand the pattern of letters on the page to mean what I take them to mean in light of my store of experience—while doubtlessly the author of those words is reaching out to me from the center of her own repertory of experience. I can't interpret her words as she does; I have no choice but to interpret them for myself, thereby using her words to write my own book within the confines of my particular life experience.

Which sounds crazy—that is, senseless in being so counter-intuitive. But is demonstrably true even so. In my case, I read Tolstoy's *War and Peace* when a freshman in high school. That is, I read the words (as translated), but what could those words mean to a young boy in upstate New York other than what he imagined them to mean on the basis of his limited experience? I didn't read the book Tolstoy wrote from his profound grasp of human relationships; I read the book I was able to understand from my naïve position on the fringe of human society. I knew what I knew, but that understanding was not up to the job of introducing me into adult society. I was a novice, a pretender, not a member in good standing, for I boasted to some of having read *War and Peace* when I hadn't understood it at all.

The great children's books work on two (or more) levels at once, that of the adult who is reading aloud to his children, and that of the children hearing the words but not caring if they don't fully grasp what they mean because exotic words are enjoyable on the limited scale of their apprentice understanding. When the children themselves come to read those same books to *their* children, they will chuckle at what they missed at first hearing. Reading books to children is a familiar situation, one where it is clear that the reader is reading to himself, while the hearers he engages are hearing for themselves. Each is fulfilling the roles of both poles of his or her situated experience, being fully lodged at the familiar and comfortable centers of the worlds around him- or herself respectively, worlds largely created by their fervent imaginations.

Voting in a democratic election illustrates the same polar construction of situated engagements in that each voter is voting *for* the candidate that appeals to her most (for whatever reason), and *against* the ones that appeal to her least, even if never confronting the candidates themselves as they are situated in their own minds and deeds. Democracy is a game to be played by mature and experienced adults, not a game for children, or those gullible to the influence of cunning persuaders. I keep hearing people explaining their votes on the basis of slogans and sound-bites, not a considered grasp of the issues. They vote for and against candidates on the basis of their own understanding cast outward onto straw men of their own making or passionate believing. Here again belief enters the picture, not knowledge, not truth, not hard-earned understanding. Passionate belief unsupported by deep and resonant experience becomes a caricature of itself if the situation you believe you face is truly a figment of your own imagination.

Which is why scientists insist on replication of research, and theologians insist on deep familiarity with and understanding of scripture. To write even a cookbook takes years of not just rattling pots and pans in the kitchen but of deep acquaintance with food, its preparation, serving, and eating. The caring has to be there, the love, the passion for making yourself happen in the right way and right place, at the right time, together with the discipline of meeting the highest standards of performance. Which is why outperforming yourself through plagiarism or by taking drugs is such a cop-out as a shortcut that violates the integrity of your own passions.

5-13. Who am I? None other than my dreamself. The one who wakes up, heart racing, from a nightmare, values and affect intact. When asleep I can't act or perceive. When awake, I can. And also interpret meanings. That's me. Same self at the core. Actor, observer, interpreter, same ageless self as when I was a kid. Doing my best to make sense of what's going on by comparing it to what I remember, figuring out what to do from the difference between now and then. Always on the lookout for new ways to be my same old self. That's my job, to be who I am.

5-14. Where am I? Another issue entirely. We're a mobile species, all over the Earth. In the jungle, on the ice, out in the desert, in the water, flying through the air, on the moon, looking through the Hubble Telescope, sending rovers to Mars, on the Web—always situated, always in our heads. Think of all the places you situate yourself in the course of one day: dreamland, contemplating yourself in the mirror, kitchen table, bus (car, train, bike), workplace, computer monitor, school, library, delicatessen, TV, Worldwide Web, grocery store, daydreams, movies, music, art, sports, ballet, book, iPad, iPod, cellphone, Twitter, YouTube, conversation, and so on. You are there! Paying attention. Being your innermost self in place after place after place because your memory is there with you. Situated, as I say.

But as I also say, that place—all those places—are in you as the personal versions you construe (construct) one after the other within your skull. The world may provide a characteristic sense of ambient energy, but you color and shape that energy to suit yourself so you get what you're looking for, and don't get what you don't want to deal with at the time. It's a matter of the polar self that you are hooking up with your other virtual, polar situation, the two coming together as the situated self you need to make yourself happen in ways that are familiar to you so you recognize yourself as the same person you've always been. Whooee, what a ride! Nobody can do it but you, for yourself. When others try to get you to hook up their way, you become somebody else. Their person, not yours. You've got standards, after all. At least habits and proclivities. A style-of-being all your own. Your own man; your own woman.

Brainwise, we're all born with way more neural circuitry than we need. We're set up to adapt to whatever conditions we are thrown into at birth. From that moment, we're on our own, tailoring our neural circuitry to the situations we engage in, discarding extraneous connections as excess luggage. Neural circuits we don't use die of boredom and neglect. Those we do use become fixed in our minds as links in our personal autobiography. We grow into our selves, shaping our minds in response to our daily adventures, those challenges that actively create the functional neural network of connections within our brains. We become who we are because of the lives that we lead; it is not possible to be anyone else. The idea behind our identity is that the fact of our survival is proof that we're doing something right, so doing more of the same has got to be the way to go in days ahead. We become creatures of the conditions that spawned us, custom-built to thrive by seeking out more of the same (*see* Reflection 5-17 *below*).

Wherever we are, we want to feel familiar to ourselves. So we seek out locales that resemble (within a certain range of tolerance) similar places we have felt comfortable in. Once a Red Sox fan, always a Red Sox fan. Fenway Park becomes a second home. We put posters of Fenway up in our bedrooms to heighten the illusion that we are on familiar territory. Or of Yankee Stadium, Wriggley Field, or whatever happens to be our ballpark of choice, just as long as we feel that's where we belong. If we picture our subjective selves in the company of wolves, we decorate our walls with wolf calendars and posters, so we feel at home in our idea of the wild, with the moon casting long shadows on the snow.

I have two maps on my bedroom wall, one a topographical map of Mount Desert Island (MDI) in Maine where I live by choice, the other a nautical chart of Frenchman Bay to the east of MDI, including Taunton Bay, my primary foothold in the universe from which to keep track of sun, moon, stars and planets making their daily rounds. That's where I watch birds, wildlife, storms, tides, boats, clouds, dragonflies, voles, shrews, hares, deer, auroral displays, and changing seasons—as fully myself. These maps are reminders of who I am, where I am. My situated self. I also have a poster on the wall, a view looking out through trees across Taunton Bay to the Bar Harbor Hills on MDI. I didn't paint the picture—Liz Dominick Cenedella did—but I might well have painted it because it captures the viewpoint from which I look out at the world—through those trees, past that ledge, to that tree-lined shore and those hills. It is no accident that that particular poster has ended up facing my bed; I carefully put it there as a depiction of my primal (if virtual) antipodes—the world I entertain in my head as a result of looking upon my unique mental situation from my preferred standpoint on this Earth.

Home decorating is home situating, creating a place to call home because it *feels like* home to us. A safe place to rest our heads and be ourselves to the max, passions included. Fixing up and decorating our home is building an extension of our own mind—that singular place where we can be ourselves in intimate detail. A home is our own turf, our territory—our personal property, not because we paid for it, but because of our personal attachment to it, to the garden, closets, entryway, kitchen, bedrooms, bathroom(s)—to the *feng shui* ("wind-water") of the entire establishment on our personal terms. Being situated is a matter of matching our passions to our environment—to our setting on this Earth. Since we are the expert on both, matching them (the two poles of our being) is our job alone. No one else can do it for us because no one else can *be there* in our stead. Their eyes, their outlook would be different from ours.

I remember one damp day in April standing at the base of a tall spruce in the middle of the woods, listening to a male robin declare himself as a true thrush, singing his lusty song to the world, broadcasting his heart and his situation to any potential mate within hearing. He lured me, for one, because he was singing my song. I understood the language and knew exactly what he meant. I was with him on every note.

5-15. Comparison I. How do we do it, find our place in this life, or as I put it, situate ourselves to our liking? That, I believe, is why we are conscious. Consciousness is not something we *have,* it is what we *are,* encompassing both poles of our being at the same time. No mean feat. In each of our cases, it happens in only one place: our respective minds. We situate our subjective selves by comparing where we *are* with where we *have been,* and use the difference to guide us in the direction of where we'd rather be.

In that, we are all helmsmen of our own minds. As I have written earlier:

> My thought is that, given the degree of consonance or dissonance compared to what I expect (am familiar with or used to), I experience a valenced signal that drives the adjustment needed to put me on the heading I desire. I steer my way by that signal much as a helmsman steers through fog by the deviance of his compass needle from his charted course. His mindfulness of that error allows him to turn the wheel to port or starboard to counter the error at each moment as he goes. In that simple image I discover the rise of William James' stream of consciousness, what others see as successive instants of working memory, and I see as my ongoing loop of conscious engagement (*Consciousness: The Book*, page 129).

Putting ourselves out there on the basis of our expectancies, hopes, fears, or desires, we navigate our way by any discrepancies we discern in our course, and act in such a way to achieve the corresponding adjustments we need to make. It's as easy, and as difficult, as that. As I see it, neural comparison of signals in adjacent cortical columns in our brains might well form the core of consciousness. Such a valenced discrepancy would alert us to the difference between where we were and where we are now, between where we are now and where we want to be. If discrepancies between images formed on the retinas of our two eyes are the basis of depth perception, differences arising from comparison between then and now situations (memory vs. perception) might well kindle consciousness itself, as the deviance of a compass reading from our charted course allows us to turn the wheel against that error in adjusting our heading. Finding our way in the world is more an art than an exact science. Much of the time we steer by intuitive dead reckoning (estimation, not measurement), or trial and error to find out what works for us. Individual consciousness is given us as a mental compass to steer by in working toward becoming the passionate persons we most hope to be in light of our feelings, values, and whatever understanding we have that can spark our creative imagination.

5-16. Growing up. I am stunned to realize how long it has taken me to reach maturity. I thought I was competent when I was five years old and could run, throw a ball, talk, and feed myself. But then I had to learn to read, to count, to add and subtract. I learned long division in sixth grade, which I considered a sure sign my apprenticeship was done. Then came the escalating challenges of seventh and eighth grades, followed by the humbling initiation into high school as a freshman. Once a senior, I thought I'd made it to the top. I had my driver's license after all, and knew a smattering of physics and calculus. Then in a flurry came college, getting married, having children, working two jobs just to keep up, changing jobs—always being sent back to Go, having to start all over again. Surely I was grown up at forty. I really mean fifty when I was in grad school. By sixty-five when I left the Park Service I figured the challenge was over and I'd come at last to the pinnacle of my abilities. But that's just when my real work began. I was no longer a replaceable module of society but was on my own, doing my thing, so had to meet my own expectations for myself. Reborn again at seventy, seventy-five, now eighty, I see I'll never get there because there's no "there" up ahead. I'm stuck in "here" and "now" because they're inside me. Everywhere I go, I bring them with me, so I'm always on the leading edge of my experience, working to transcend myself, but finding my expectations and yearnings getting farther ahead of my aging frame.

That is the life story of the situated self. The two poles of mental life, the inner self and outer virtual situation, grow apace on comparable levels of experience. As the self grows, its world grows; there's no catching up. No acme of accomplishment. Not even a plateau where you can sit back, smell the flowers, and enjoy the view. Every new level of understanding and attainment opens onto an increasingly complex and puzzling view of the world. Questions, nothing but questions. Because the disparity between memory and perception persists, there's always an error signal nagging me to get back to my true heading, which I struggle to do, but I keep misjudging winds and crosscurrents, so have to keep making adjustments, learning as I go.

Learning by doing—trial and error—that is the name of the game we have to play just to survive. Nobody gets it right all the time because everything is changing in different directions at different rates so our equations are always out of date, and there's no remedy but making another trial in the hope that this time we'll get it right, which we won't, but we have to try just the same.

If our journeys and struggles always lead back to the same beginning, what's the point of doing anything at all? Why not just scream, "I can't go on this way!" and call it quits? That is the ultimate demonstration that consciousness—as an agent of individual survival—is an arm of natural selection. If we don't live by our wits, we die, leaving fewer descendants than if we keep striving to the end. Consciousness is how we adapt to whatever situation we put or find ourselves in. It makes each of us precision (or blunt) instruments of our own reproductive success in an ever-changing world of tsunamis, hurricanes, volcanic eruptions, earthquakes, wars, revolutions, train wrecks, pandemics, crop failures, cancers, accidents, economic collapses, climate changes, betrayals, exploding stars, crashing asteroids, forlorn hopes, and all the rest.

Nobody said it would be easy. But what we find as we grow is that full, deliberate, conscious consideration is the best tool we have if we want to deal effectively with whatever situation we find ourselves in when we wake up to the world of tomorrow.

5-17. Identity. When we wake up to that world of tomorrow, who is it exactly who wakes up? If I am the dreamer, then I am also the waker. It is I who wake up. Who then is I? I am I. Who is that second I, then? You can't define something in terms of itself. I'm searching for an external point of reference. There is no external point of reference I can point to in defining myself. As I said, I am I, or if you wish, I am me, both subject and object of myself; end of discussion. Please, no more questions. I am, I am, I am. I am a matter of declarative assertion. I am that I am.

More often, when asked who we are, we rattle off our name. If pressed, we say where we live, or offer a smattering of personal history such as where we were born, who our parents are, where we went to school, and so on, in order to help others give us a place—a situation—in their system of reference to help them recognize us as familiar the next time we meet. Our identities place us in situations where we are apt to be met, so they contain a *from* element associating us with the doctor's office, say, the school, the library, the street in front of our house. I often introduce myself as "Steve from Planet Earth," which I offer as my true place of origin. People seem to remember that, even if we haven't met for a year. Oh, yes, that Steve, the one from Planet Earth. Our identity is often more a peg for others to hang our physiognomy on than a précis of whom we might be.

Another sort of identity is association to something we have done or achieved (Nobel laureate in physics), to someone we are related or connected to (Othello's wife), or association with a group we are a member of ("red" George as opposed to "black" George, "tall" George, "gorgeous" George, or others sharing the same first name).

More commonly, an identity reveals how we think of ourselves or others think of us. That is, how we are situated in our own or someone else's mind. "Sometimes I feel like a motherless child, a long way from home." "I'm just a lone wayfaring stranger. ""You're the cream in my coffee, the salt in my stew." These come from lyrics sung by or about situated selves. If we're not situated in some way we tend not to sing, or attract others who do. We wake up to the clothes we took off when we went to bed. Those clothes tell us who we were yesterday, and give us the option of being much the same person today. Each morning we slip into our lives as told by the furnishing of our personal space. We wear the clothes we find on the chair, eat the food in the fridge, drive the car in the driveway, open doors we have keys to, remember what's on our minds, and we're off—noticing, interpreting, doing—in a new day.

5-18. Understanding. The presence of memory is made evident in perception by our ability to recognize and categorize sensory patterns we have seen (heard, tasted, smelled, touched, etc.) before. I visualize recognition as reactivation of portions of neural networks established through repeated use or in the presence of strong feelings. We use the established network to *understand* the pattern in light of prior experience. In that sense, our network stands under and supports our present grasp of the pattern we are sensing, placing it within a field of related patterns each having particular characteristics and their associated verbal labels. That orderly arrangement of related patterns and labels is our means for reaching out to and understanding our experience in light of relationships established by the specific layout of those patterns in memory. If we can't recognize a sensory pattern, we are unable to place it within our field (network) of prior sensory experience. In that case, we say the pattern has no *meaning* for us in that we are unable to assimilate it to our existing repertory of identifiable categories by some combination of name, shape, color, motion, texture, size, scent, taste, feel, or other familiar qualities of experience.

When we reach out to others, we reach from our understanding of the situation that binds us together so that we share similar or overlapping aspects of that situation in common. But what we exchange are patterns of energy, hopefully—if we speak the same language—recognizable ones that have meaning in our respective lexicons of labeled patterns of experience. But meanings do not easily leap from one mind to another without support from a common background of experience. Meaning, that is, is a property of *experience,* not language itself, not simply making and understanding the same sounds and gestures. So if I study zoology and you study zoology, that does not guarantee we have the same understanding of what zoology *is.* Our specific courses of study are different in each case, rooted in different fields of understanding because our respective fields of life experience in and out of the classroom are bound to be different, even though we sit right next to each other in class. With the result that your study leads you to become, say, a marine ecologist, while I go into fruit fly genetics.

The same stream of incoming sensory *data* might well have profoundly different effects in interacting with the fields of understanding of people having divergent backgrounds of experience. Data, that is, doesn't become *information* until it interacts with an individual's unique field of understanding. And then it might well turn out to be different from the information in the mind of someone else providing a different context of understanding for the same sensory pattern or "data."

We are situated in respective fields of understanding that often turn out to be more *experiential* than cognitive or logical. If we undergo a disciplined education in systematic thinking, we rationalize or omit the incoherent parts, so have gaps in our understanding just as we would from a more ad hoc education in art, say, filmmaking, or home economics. Invariably, we know what we know, and don't what we don't. We carry the worlds we are situated in with us everywhere we go, and face the life-long challenge of expanding them at every moment until the very end. When we die, we die twice-over: primarily to our selves, but to the virtual worlds of understanding within which we situate ourselves as well.

Conversely, when we are born, we are born twice-over, to our unique selves and to the progression of situations we move through in living a life. We are born to remember the passionate fitting of ourselves to our current situations so that, when we face into new situations, we have precedents to build on. In that sense, we build ourselves inside-out on the basis of what we remember and recognize as familiar. This fits into my world because it reaffirms who I am; this does not fit because it disregards—and so threatens—my core self. Imagine what education could be if it nurtured the situated self in each child's heart instead of paving over that child's hopes and aspirations with other people's notions about who they want her to be.

5-19. Reason. Just to be clear, I do not consider myself a rational being. I am neither systematic nor logical in my approach to solving problems, but more intuitive and experimental. To me, reason is more a matter of being earnest, of applying myself, than of adopting one particular method or another. I use "reason" in a loose association to whatever makes sense to me in a given situation. That is, reason and rationality apply to whatever I (my situated self) am able to bring to particular occasions. Reason is not so much a systematic virtue as it is a distinctively personal quality of mind suggestive of how I rationalize my understanding to myself. In that light, situated selves are inherently rational in taking situations themselves into account as the momentary and subjective context of consciousness. That is, I use "reason" to signify the quality of being self-aware. "The dream of reason," then, leads to becoming helpless, anxious, or overly fearful (*see* front cover).

5-20. Creative imagination. The beautiful thing about being situated in our own understanding of ourselves and our worlds is that there is no law requiring us to limit our actions to precisely those hard-won understandings. We can *improvise,* turning out fiction as well as guidebooks and academic treatises, writing plays, choreographing dances, turning Dante's mind into a graphic depiction of his thought—as long as we remain true to the passionate *sense* of what we understand rather than the literal or experiential incidents by which we developed that sense. We can transcend our own limited thinking to walk among angels, as Mohammed did in writing the *Qur 'an,* Dante the *Divine Comedy,* Shakespeare his plays, Stravinsky in scoring *Rite of Spring,* Beethoven his string quartets, Franz Kline in spreading black marks across a white ground—all expressing themselves figuratively, symbolically, metaphorically, passionately, expressively—while remaining true to the spirit of their situated understanding. Saying what they want to say; getting their message across. To me, this is proof of the experiential nature of our situated understanding, which through creative imagination, can give rise to expressions of that understanding without being confined to a literal representation or transcription of the experiences within which it originates.

This tells me a lot about the nature of experience. We can channel the passion, authenticity, and coherence of our situated selves into a variety of media. We are not stuck with the same old, same old, but can transcend ourselves in reaching for higher expressions of our most passionate selves. Once we realize a situation on a passionate level, we can transpose that situation into a different medium of self-understanding and expression, remaining true to our passions in the process.

5-21. Meaning I. The situated self is the department of mind concerned with meaning. That is, with fitting sensory patterns to preexisting fields of understanding as a basis for considering appropriate actions to perform in response. Poised for action, the self appraises sensory patterns for their relevance to prior experience. If it does not recognize a pattern, it can fit it into the closest category of understanding, or if that seems too forced, it can create a new category (which takes emotional commitment or pronounced rehearsal and repetition). If that is too much work, the pattern can be dismissed as too trivial to bother about, which is how many people deal with changing scenes they have trouble keeping up with. It takes passion to welcome a novel pattern, and passion to reject one; positive passion, negative passion, thumbs up or thumbs down. We all form attitudes toward familiar and unfamiliar experiences, either turning toward or away from potential engagements based on their degree of familiarity or strangeness. We remain open to some experiences, become closed to others. The big question is: What does it mean (if anything) to me now where I am? Meaning, like self, is situated at the heart of consciousness in time and space—in recognition of sensory impressions and realization of desirable actions. It is the link that joins the two so that action meaningfully *flows* from perception. Meaning is *the passion of the life force in process of fulfillment.*

5-22. Phenomenology. The –ology (legend or logic) of phenomena is that sensory impressions (appearances) are not meaningful on their own but take on meaning by being fitted to the understanding of particular persons with unique histories of experience (via memory). By deliberately holding back that meaning, the tendency or passion *to make meaning* can be made conscious, and so claimed as the responsibility of the perceiver rather than being a property of phenomena themselves. Once that responsibility is claimed, the world is transformed into patterns of energy impinging on our bodily senses, and the perceiver is elevated to the position of meaning-maker-in-chief. That is, the one who receives the message is in charge of interpreting it, every bit as much as the sender must understand it for herself. Which puts accountability for meaning (and responsive action) on both sender *and* receiver, writer and reader, not one *or* the other. Once the implications of phenomenology are grasped, the world (including this manuscript by way of example) begins to make sense. No aspect of understanding has more profound implications than can be discovered through adoption of full responsibility for making sense of the world and the impressions it prompts in conscious minds. The scales fall from our eyes. We see the light. If there is madness in the world, it is likely to be our individual affair, not collectively the world's.

5-23. Memory I. Though I'm uncomfortable with the computer analogy, like IBM's Deep Blue in defeating Garry Kasparov, we are remembering machines, applying the contents of our training (programming) to every situation, predicting outcomes, selecting the course of action that, we feel, has the best probability of success. The similarity is not in our computing skills, but in our capacity to bring the past to bear on the present—and on the future. In our case, the past as we remember it, not the way it was. If we had no ability to remember, consciousness would play very differently. We'd live in a sea of meaningless impressions—this, then this, then this. We'd not be long for this or any other world because we couldn't learn through study or trial-and-error. Instinct would be our guide, reducing us to the level of insects with a repertory of very few automatic tricks (routines). We'd shrink in size because we wouldn't have the smarts to get all the calories we'd need to maintain our bodies. When bighorn sheep were reintroduced to the Grand Tetons, they lacked the memory of how to get to their winter feeding grounds, so perished even though well-suited to the terrain. Without memory, we'd not only be unconscious, we'd be dead.

We are who we remember to be. Which explains the inertia of the past, and why it takes a younger generation to turn the old guard out of its entrenchments.

5-24. Numbers. Numbers and the operations we perform on them open doors onto the comparisons and relationships at the heart of personal consciousness. For example, on a scale of one to ten, I might say I agree with you at about the seven level, with ten standing for total concurrence. We have units set along a scale, constituting a certain range, and we measure (compare, gauge) a certain relationship against some such scale in trying to tell others what it feels like to be us. Numbers are like words in being cultural tools for quantifying experience in order to share it with others.

Numbers aren't based on objects in the world because they are experiential or phenomenological in nature, the products of comparisons, relationships, and estimations within our minds. We learn to count on our fingers, uttering a particular sound as we point to each one *as if* we were counting our external fingers and not our internal gestures and utterances. Behind numbers lie the experiences they represent, the comparisons, proportions, relationships, ratios, equations, valences, similarities, differences, correlations, units, positions, changes, among other dimensions of experience. Calibrated by cultural agreement on what units are thought meaningful, numbers measure or gauge the degree to which something is present in awareness, and how it measures up in relation to something else. Degrees of change, difference, sameness, or presence in awareness are the forte of numbers.

My forte is what those particular degrees mean to me. I don't view numbers as being meaningful in themselves, but as indicators having implications over and against my fears and expectancies. It is those qualitative implications that concern me. A body temperature of 102.7 degrees Fahrenheit by itself means nothing; its meaning must be gauged against an everyday norm (such as an oral norm of 98.2° ± 0.7°F). Even gauged to the nearest tenth of a degree, a given temperature takes on meaning only when placed against a background of expected daily temperature variations. Any such background is not inherent in a number itself but is derived from a background of experience contained in memory. It is the qualitative disparity between performance and expectation in a particular mind that spurs consciousness, the qualitative similarity that triggers recognition.

5-25. Speech II. When it comes to our self-proclaimed species binomial *Homo sapiens*, I'd say we are more *wordy* than *wise*. Sounds flow endlessly from our mouths; whether they make sense or not is another issue because it depends on the situation we are addressing, that is, from which our words arise. For, indeed, that is where words come from, the particular situations we are in (or thinking of) at the time we open our mouths. To discover what we mean, we must consider how we are situated when we make the sounds that we do (*see* Reflection 1-16; Figure 2-5.1).

Invariably, our words issue from the deep structure of our minds where we are situated when we feel the urge to speak. It is our situated selves who do the talking, converting (transforming) the tensions and relationships inherent in thoughts and impressions into passionate speech, giving voice to the meanings they recognize in the clash between what they remember and what they experience at that moment. Noam Chomsky gave up his notion of transformational grammar, but with *deep structure*, I think he may have been on the right track. Words are passionate expressions of tensions giving structure to situations arising in current experience as we strive to understand it. It is our deep (not our superficial) selves who do the talking, modulating the column of air rising deep in our throats. By the time teeth, tongue, and lips have their say, the meaning has been long shaped in our depths.

When I am interrupted in the middle of paying attention to something, I often utter an expletive that speaks to my discomfort at being caught unaware—venting a primal sound, not a word. Its meaning comes prepackaged in that sound because it's part of the repertory of sounds I draw on in such events, depending to a degree on the company I'm in. Too, when something nice happens that reinforces my attention, the surprise or humor (positive disparity) I feel when things go better than I expect is expressed in a smoother sound of satisfaction ("ahhh," "yum," "mmmm," "wow," "good"). Addressing an audience, I strive to share how it is going with me (with my situated self), my words choosing themselves to echo the deep structure of the process I use to make meaning by passionately fitting my current thoughts to the situation on which my remarks are focused at the time. When I part my lips to speak, the words are already there, sent up on the dumb-waiter from the deep, experiential kitchen where my meanings are prepared in more-or-less digestible form.

Another way to put that is we are moved to say *what's on our minds.* In this chapter I've been trying to put down what that might be: understanding our sensory impressions in meaningful ways in order to decide what to do about them within the situation we believe ourselves to be in. Which sounds dry, but in felt experience is based on the caring, love, and passion that drive (motivate) us to do what we do, including speaking our minds. So do we rise to the occasion because nothing less than the meaning of this moment in our lives—the very turning point between past and future, this instant of awareness—is at stake.

What else can we do but speak the story of our fears and desires as they bear on our survival?

I hear a voice saying, Speak for yourself, Steve. Well, then, I'll let you in on my secret. I care about nothing more than the subject matter I'm dealing with here, the situated self. *My* self as it is situated in this moment near the end of my life. Here I am, trying once more to get clear on the flow of conscious activities through my mind, much as I have been self-reflecting ever since that day in June, 1934 (when I was 21 months old), playing with a big dog under the dining room table, standing on a chair, climbing onto the table, being lifted up to peer down into the white basket containing a tiny baby, who, it turned out, was to play the part of my little brother. That was the first moment of self-awareness I can remember, of *being there* as myself apart from any experience I had of the subjective universe I was then gathering about myself and beginning to engage.

I don't remember making any sounds then, but I now realize I was becoming proficient in four languages: 1) the language of action in the world, 2) the language of energy relayed as sensory impressions, 3) the conceptual language of recognizable patterns of sound as uttered in suitable situations by my family and other cultural agents, and 4) the emotional language underscoring the whole event as worthy of remembrance. I don't recall the sound "baby" being uttered by anyone, but I'll bet it was, and that was the label I associated with being lifted up, seeing a person smaller than myself, forming the rudimentary category I would sort similar patterns into ever after, together with the emotional kicker that bound the incident together as one of the earliest entries in what has become my autobiographical memory.

Producing fast results at slight expenditure of energy, speech is an efficient social substitute for slower and more costly commitments to physical action. As such, it can be separated from direct action, its component parts conveniently stored in semantic memory, to be brought up at another time in reflecting on prior events, or planning future activities. By definition, speech is a culturally administered social activity serving as a medium of interpersonal exchange and engagement. Specific venues are set aside for certain kinds of speech and related activities: baseball talk, boxing talk, God talk, baby talk, sex talk, money talk, war talk, food talk, clothing talk, and so on. The situations apt to be met in such places provide the underpinnings for whatever speech is likely to ensue in such settings. Culture provides the occasion and vocabulary; as individuals, we supply the situations and passions those places elicit in our experience, and the utterances that rise within us as evoked by the inherent structure and tensions we discover in those situations.

Because of our double duty in being true to both poles of our situated selves, inner and outer (subjective and objective, actor and perceiver)—in the presence of speech, we can imaginatively balance the complementary roles of both subject and object, modifying them for suitable emphasis on the level of detail we are concerned with, linking them by choice of an appropriate verb to clarify their relationship as we choose to present or understand it at that moment. As I put it in *Consciousness: The Book:*

> Speech acts, I suggest, spring from a situation as a kind of proposition or assertion characterizing some salient aspect of that situation as depicted within a particular loop of engagement—likely to be heard in other minds from a variety of different perspectives. That is, speech flows from our ability to categorize sounds as meaningful in some way. The sounds are interpreted in speakers' and hearers' minds through loops of engagement by which those sounds summon (invoke) conceptual interpretations while, simultaneously, those same interpretations affirm such sounds as making an appropriate fit to the situation at issue as it is construed in particular minds (page 155).

Speech is propositional in nature, an effort to present the structure of, and tensions within, a particular situation in verbal terms. Our reference is not to the world but to that situation as we experience it at the time. Continuing:

> I am saying, basically, that speech acts announce the essential workings of our minds by drawing attention to events as salient aspects of felt situations. When we speak, we literally speak our minds, and when we hear, we have the words speak to us within our current situation as influenced by prior experience. Listening to speech sounds, that is, we act as a kind of ventriloquist projecting our meanings onto the moving lips of the speaker. Speech is situational in nature, and the relevant situations are invariably internal to the loops of engagement maintained by the particular people involved. Which suggests that in speaking our minds, we can take only one detail at a time, leaving out much that could be said. We speak our minds incrementally and sequentially, making our point, then moving to the next, dealing with our felt situations bit-by-bit, parsing them into details one after another as we choose to emphasize them and relate them to our overall understanding as we are able to articulate it point by point. Thus we build sentences by predicating (asserting) something about a chosen topic or subject, adding modifiers, details, and emphases as we go, all in good order (sequence) so we will be understood by members of our speech community, those who taught us to speak in the first place (same source).

For me, there is no speaking about speech without making reference to the complementary loops of engagement within which it takes place. Speech is but one form of human engagement, right up there with eating meals together, playing together, working together, learning together, loving together, surviving together. Bringing me to the brink of chapter six, which deals with Loops of Engagement. ○

Chapter Six

LOOPS of ENGAGEMENT

6-1. Metaphors. Turn-by-turn, my conscious mind spirals like a helix or thread of a screw set to a certain pitch, pulling me into the future around a constant pole of awareness centered on my situated self, revealing a passing vista upon virtual worlds as the endless, ramping path of my personal travels through life. It has taken me thirty-five years to propose such an image as a model of my—and possibly your—streaming consciousness, the process by which we reinvent ourselves again and again by dizzily thrusting along the autobiographical timelines of our respective rounds of lived experience. We bore through life like a drill bit turning, turning, advancing deeper as we go. That is where we live, on that cutting edge of our mortal coil, the life force—as long as it lasts in our particular case—thrusting us ahead, round after round, ramped cycle after cycle, enabling us to learn by trial and error how to cling to the advancing edge of our own possibility as it cuts its unique yet continuous path. So do we get ahead in life, as Earth gets ahead in the solar system by centering itself on the sun and its multi-dimensional journey through the Milky Way and larger universe. As our minds get ahead by noting and comparing shifts in awareness at every moment.

I intend that paragraph to capture some of the flavor of what I mean when I refer to loops of engagement in trying to get at the motive force behind consciousness, the thrust that brings our days to us, and us to our days, enabling us to live a life of engagement while maintaining the integrity of one mind in one body. You won't find neuroscientists writing such a paragraph because they are locked into the imagery of the brain as a computer or information processor. But I am looking from a different perspective, backed by a different stock of assumptions, so I view consciousness as it presents itself to me without having to see through a set curriculum of neurons, synapses, and neurochemicals. The metaphors that occur to me are in terms of helices, screws, drill bits, propellers, spiral staircases, and stitches such as a seamstress would make in joining pieces of fabric together, as I see my inner and outer worlds stitched moment-by-moment to form the quilt of my felt experience. I can't explain my own consciousness; the best I can do is describe it as I discover it to be through self-reflection, a discipline within consciousness studies every bit as worthy as philosophy, psychology, and neuroscience, and without which consciousness would remain a mystery forever because we would not recognize it when we met it head-on as we do every day.

6-2. Native bent for engagement. Questions prompt engagement, so I picture consciousness in terms of three successive queries: 1) What's going on?, 2) What does it mean to me?, and 3) What should I do? Taken together, perception, meaning, and action make up the plot of much of my conscious activity, which I watch with interest as a play in three acts. Even my dreams fit that broad scenario, as do many of my social engagements, studies, recreations, projects, relationships, and trains of thought. All underscored by feelings, without which I probably would not be moved to do much of anything. It's not that life imitates art so much as that art expresses life. Our inner life is the model for cultural events. Because we survive through concrete engagements, art excels at attracting and holding our attention. To call dreams "theatrical" is to miss the point: theater is meant to hold our attention, so itself is "dreamsical" in being based on our native bent for engagement.

What sets me going on a particular round of awareness is often noticing a disparity between what I hope for or expect and what actually happens. If everything goes as planned, there's little need to put myself out by paying conscious attention. But if a question or issue arises, I had better get on it to find out how serious it is, what it means, and what I had best do. Any disparity in my expectations, whether positive or negative, is an invitation to engage the situation with the aim of discovering what needs attention. As I have written, the survival prize is won by the vigilant, not those who are too tired or busy to care.

6-3. Salience. What gets my attention? Sunsets, moonrises, gusts of wind, claps of thunder, the call of a crow, sight of a snake crossing my path, fire sirens, loud voices, sudden movement, bright colors, alarms going off, novelty, melodies, beauty, ugliness, lingering tastes, strong smells, sudden changes in temperature, and so on. These are events declared by patterns of energy in the virtual world. But, too, salient inner events also get my attention: headaches, toothaches, hunger, worry, fear, anger, yearning, caring, love, trains of thought, fond memories, nightmares, erotic urges, flitting ideas, plans, relationships, harmony, and dissonance, to name but a few. It makes little difference to me if my mind is kindled within or without. The point is engagement, not who starts it, as the point of my beating heart is that it keeps going, and the point of my breathing is exchanging stale for fresh air. Without pumping heart, heaving lungs, reaching mind, I am nobody, nowhere, no time.

As I see it, mind begins with a mental comparison. To give three examples, that comparison might be 1) between two mental states such as fear and hope, dread and desire, expectancy and accomplishment; 2) between what is remembered and what is currently experienced; or 3) between action toward a goal and the effect it produces. Such comparisons are no flight of fancy because we know the disparity between comparable signals from our two eyes contributes to depth perception, and sensory feedback helps the cerebellum achieve smooth fine-motor control. Adjacent cortical columns, for example, offer a venue for gauging the similarity and difference between comparable signals, leading to the possibility that a valenced adjustment be made one way or the other. I see my own mind as just such a *comparator* between mental states, juxtaposing pro and con, good and bad, yea and nay, desirable and undesirable, weighing the advantages and disadvantages of each, and coming to a decision regarding which is preferable in a given situation.

Eureka!, consciousness. Loops of engagement provide threads of awareness based on successive rounds of comparison as measures of progress. Commotions and alarms—like events in Tiananmen Square, Tahrir Square, and Zuccotti Park—are trying to tell us something by gaining access to our notice that we might have missed up to now.

6-4. Comparison reveals change. Comparisons between successive rounds of engagement give us not only consciousness, but our inherent senses of time and space. Comparisons produce awareness of relative changes or discrepancies. If they are not our doing but come to us through the energies we detect on the far side of our senses while we remain still, then we credit such differences to the flow of events beyond ourselves, and call them changes over *time*. If, on the other hand, we are active while detecting such discrepancies, then to be clear about it, we need to subtract our own activity from such a result, that adjustment producing a rearrangement of our vista in three-dimensional *space*.

When, in Sarasota after World War II, I saw a train coming toward me through the swamp as I was walking as fast as I could toward the train (*see* Reflection 5-9 *above*), my mind had to distinguish changes due to my forward motion in space along the tracks from changes due to the train growing larger with time in order to estimate both the time and place of our mutual collision, which hopefully would allow me to reach the trestle and slip beneath it before that might happen. I wasn't aware of computing that point of intersection, but I had an intuition that it would be close and I could make it across that distance in time to beat the train, sparing me having to throw myself into the swamp on either side, with regrettable loss of my loot from the airfield. The scene is etched in my memory to this day as a battle in time and space which I happened to win, one of the milestones along my life's journey through space-time.

I passed a somewhat similar milestone in driving across the country with my family when we moved to Seattle from upstate New York, and I was eager to get my first glimpse of the Rocky Mountains (*see* Reflection 3-3 *above*). Here, more fully, is how I processed that event in terms that are perhaps more meaningful now than when I introduced it in that earlier reflection:

> When driving west across eastern Colorado with my family, I expected to see mountains ahead, but inquiring with motivated eyes, I was answered only by a row of low, phenomenal clouds, with not one mountain in sight. The place was right to see mountains, but the time was not right for me to see them because it was late August and the Rockies were covered with snow, while I was expecting to see bare, summer mountains. The snow appeared in my eyes to be a line of low clouds floating above their own shadows, which were actually the mountains below the snow line—the very ones I was seeking but could not see. Because I resisted seeing snow, my loop of engagement was broken by my demand to see mountains without snow, and what I saw was clouds that made a better fit with what I thought I wanted to see. I was almost right, I was convinced I was right, but in terms of that particular situation, I was dead wrong. Expectations wrought in the past were no match for my current situation, so my looping engagement with the landscape before my eyes was troubled and I simply would not let those mountains form in my awareness. The harder I tried to make them fit my expectations of how mountains should look, the better clouds I made of the snow that covered them. Despite my best efforts, I forced those clouds onto the mountains I was looking for, hiding them from view. The harder I looked ahead, the less I could let them appear. In such cases, the past rules the present every time. Until in this case I noticed the clouds stayed

exactly the same for half an hour—which no self-respec-
ting clouds would ever do. This was a new discovery,
changing everything. If they weren't in fact clouds,
what could they be? Reframing the question from *where*
are they? to *what* are they?, I was answered with snow-
capped peaks—the Rockies, indeed, right where they
should be, right where they had been all along.

This tale tells the difference between assimilating
the world to preconceived expectations, and accommo-
dating expectations to the world that is there. The landscape that I saw reflected past situations I had been in.
My outlook was rooted in my earlier experience, which
I cast ahead of me into the void of the unknowable-in-
itself world. And then in a flash I updated myself, never
to forget what I learned on that day in that place—how
the snow-covered Rockies actually appear in late August as phenomena to be interpreted in my mind, which
was all new to my then thirteen-year-old self (*Consciousness: The Book*, pages 125-126).

Here, then, is a tale of failed comparison because my standard for winter snow coming in November was not up to gauging the arrival of snow on the Rockies in August. As a result, I
dutifully turned snow into clouds because I could not outrun my
limited experience in growing into a larger understanding of the
energies of the world. I was stuck. Stuck in my expectancies,
which were firmly rooted in a past I was familiar with. As a
result, I was not competent to witness the Rockies on their terms
in their time and place. I was stuck in my head, my cortical columns unable to perform their comparisons with meaningful
results. I suffered from a bad case of mind over matter. As many
members in Congress were stuck in not accepting that Barack
Obama won the presidency in 2012, so they dug in their heels
and refused to help govern the nation, the job voters elected
them to perform, unwilling to bring their past hopes into line
with current events. When their pledge not to raise taxes expired

with that election year, they were freed to do the job they had been elected to do. It's the same as not growing up, as for half an hour I dug in my heels and unwittingly refused to let Colorado be different from New York.

The art of consciousness is in dealing with changes effectively, wherever they arise, inside or outside, or both at once. In moving ahead, not holding back. Which is harder to do than we commonly assume. Jet lag is a sure sign our bodies have trouble keeping up with our plans and expectancies. The sinking of the Titanic was due to the builder's unwillingness to imagine the true bulk of icebergs in the North Atlantic in relation to his fragile ship, double-hulled though it may have been. Holding fast to outdated expectations is like believing your enemies are as puny as the vermin you imagine them to be. We underestimated the Japanese in World War II, the North Koreans, the Vietnamese, the Sunni-Shia divide in Iraq, Al Qaeda and Taliban in Afghanistan. Military planning is always based on prior conflicts, so is perennially out-of-date. It takes work and attention to keep up with changing selves in changing times. That's why, to some, creationism is more reassuring than evolution, sticking to basic principles more comforting than accepting change as a fact of daily life.

Personal consciousness does the best it can under any and all circumstances. We fail when consciousness can't gauge self-changes and it-changes correctly for no other reason than our minds 'r us. We don't *have* minds, we *are* minds. All minds are fallible in dealing with change because of necessity they are based on neural networks shaped by events in the past, not the present. The more solidly they are built, the less likely they are to deal with the next big thing when it comes, which it surely will in the form of a tsunami, revolution, violent confrontation, population explosion, famine, flood, drought, or other change we have failed to anticipate and prepare for. We'd rather global warming would just go away because it calls for so many drastic adjustments to our daily routines. We prefer to ignore it than deal with the consequences, which tells us more about our minds than the situation we confront. The same is true of the burgeoning takeover of the Earth by its human population, and of the inevitable struggle by developing countries for parity with the most consumptive economies among us. Developing no longer, they have now become us.

6-5. One step at a time. Not much of a skateboarder, skier, swimmer, dancer, or runner, I still prefer walking to other forms of human locomotion. In fact I see walking as the trait that defines us as human. It's not our minds so much, our humor, or dependence on fire, but putting one foot in front of the other that makes us who we are, go-getters of the world. And walking demonstrates how naturally inclined we are to progress through successive rounds of our loops of engagement, placing our weight on one foot at a time, stretching with the opposite leg, swinging ahead, placing our weight on the other foot, stretching, swinging, stepping, stretching, swinging, stepping—ahead, always ahead. Which sounds like a strategy based on alternation, not continuous rounds of engagement. Except that at each step our perspective on the world is one step ahead of where we just were, so our sensory impressions must keep pace with our feet in order to gauge where to take the next step, and the one after that. Our experience of walking depends on fast-moving changes in both action and perception in a continuous sweep through our minds, not just a tick-tock rhythm of legs that runs free of our minds altogether. Step by step, our minds and our feet move ahead through round after round of consciousness which streams through us as we go, more like a flowing river than the swings of a pendulum.

Not that we have to think about moving our legs once we have learned how to walk, but we do have to watch where we're going at the same time we are striding ahead. Potholes, cracks in the sidewalk, and random piles of dog shit remind us to be wary as we go, to stay alert to our surroundings, and not totally lose ourselves in thought about other matters. But I find I do manage to think as I walk, better than I do sitting in a chair without that steady rhythm to whet my mental activity. Thinking, I find, goes with walking. If I get stuck while writing, I go for a walk to stretch my legs and my mind at the same time. I always keep working on that next sentence I cut short, and rush back to write it down before I lose it again. Walking, as I see it, is an extension of my mental activity, or vice versa, the two complementing each other, coordinating the rhythm of my legs with that of my mind. If I were to become a professional thinker, I would join the union of peripatetic philosophers who talk as they walk back and forth in their sheltered but drafty colonnades.

The rhythm of walking is similar to the rhythms of music in repeating the same beat over and over again. Think marching bands and funerals in New Orleans. Think dancing, sewing, knitting, working on an assembly line, doing the same moves with subtle variations, while always alert to the differences at every instant. These are not merely repetitive activities like striding on a treadmill; they get you somewhere so that the beginning and ending are not the same. They are helical engagements that move you to a different place in your experience, whether performing in a jazz ensemble, dancing with your partner, making a potholder, knitting mittens, or assembling a computer. You have something to show for your effort. You have been somewhere, given a performance, created something that didn't exist before. Made good use of your time, even created time itself, for which you may or may not be paid, but you have enriched your experience and lived a stretch of your life in a particular engagement. As I—my eyes, my thoughts, and my fingers—have tapped out this paragraph, and I can honestly say, now I'm getting somewhere, which is not where I was two minutes ago.

6-6. Channels of engagement. I stress the serial nature of our engagements, step after step after step. But, too, I visualize channels of simultaneous action by which we engage in different ways at the same time. Rodin's sculpted *Thinker* could be shown scratching his chin or the back of his neck. Smoking is often a

126

side effect of our social engagements, as are munching chips and drinking beer. Sometimes we hum to ourselves, tap our toes, or dance lightly with our feet when otherwise engaged. Most people gesticulate as they speak. In conversations, we often mimic the posture of the person whom we are engaging. Or send conflicting messages by not looking at, or turning away from, our audience. Parallel channels convey a sense of others' intentions if we are adept at reading such signs.

6-7. Born to engage. Are things as they should be?, that is the question. Instant by instant, always the question. And, frequently, the answer is no, so we rearrange the furniture, clean the house, sharpen the pencil, dust the room, straighten our papers, pay the bills, mow the lawn, weed the garden, pick up the kids, edit the manuscript, wash the car, and on and on, engaging our surroundings, getting them into shape, then letting them slide all over again—until we ask one more time: Are things as they should be?

Over four years, I wrote *Consciousness: The Book* in nine drafts under three different titles, and now am writing a new work based on the next generation of my thinking about consciousness. I've been studying my mind for thirty years, and will keep going until I don't have any more secrets or any mind left to write about. Start to finish, my life is turning out to be one long series of engagements: going to school, raising a family, working at thirteen different jobs, serving two years in the Army as a Signal Corps photographer—every day spent engaged with the project(s) du jour, leaving behind me a paper trail of photographic prints, articles, books, manuscripts, lesson plans, graded exams, idle jottings, to-do lists, letters, emails, over 500 posts to my blogs—not to mention the eighty-eight PowerPoint presentations on the hard drive of the computer I'm working on at this moment. Those engagements collectively tell the story of my life in gross detail. The finer points are told by how I spent each second of concentrated attention, now thankfully forgotten.

My view is that we are *born to engage* because that is how our minds work in cycles of reaching-out to the world, world energy reaching-in to us, in a series of helical engagements lasting variously for less than a second up to multiple decades. Engagements are the basic units of our days as we live out the stories we tell ourselves in discovering meaning in what we do. An instant without meaning is no instant at all—it doesn't exist. We cannot live without passion and meaning, and engagements are our way of achieving them through our own efforts to make ourselves happen in the world in deliberate and fulfilling ways. As time and space exist within us, subjective passion and meaning are discoverable in life only by looking within.

6-8. Cultural displays of engagement. A grand clearinghouse for getting people to engage one another in mutually agreeable ways for the common good, *culture* is the context we create for ourselves by each doing his or her thing in concert with, and out of respect for, those around us. We contribute our personal passions and meanings to our culture by living as we do, being sensitive to the ambient of world energy that surrounds us, while acting to contribute to that ambient by adding our personal force to the collective force of the world. But first we have to grow into the multiple layers of our cultural heritage—family, language group, community, religion, economic system, and so on. We learn to be ourselves by interacting with others through engagements they initiate as they introduce us to their various ways of being in the presence of infants and children. If they are successful in feeding and caring for us, we learn to meet other challenges by mimicking or performing variation upon the styles we are exposed to, learning to engage by interacting in increasingly complex and powerful ways according to the passions and meanings we discover in our growing selves.

Learning to walk, talk, play, dress and feed ourselves more-or-less in accordance with the culture we are born to, we gain confidence in becoming ourselves through membership in that particular culture, whose ways become ours for life, always subject to the personal abilities, preferences, attitudes, and styles by which our situated selves engage with and make us happen in the world. We learn to engage through being engaged by others, in the process becoming our acculturated selves. Engagement is at the core of our being; without it, we would have to reinvent what it means to be human on the basis of our infantile experience, not the cumulative experience of our ancestors.

Culture, then, becomes a kind of museum of possible ways of making ourselves happen in the world: musical ways, religious ways, political ways, military ways, athletic ways, instructional ways, recreational ways, family ways, creative ways. From that cultural repertory, we pick and choose the ways that channel our energies, passions, and meanings most effectively within the cultural time and place we happen to occupy in our own minds. Culture, that is, sets the range of options we as unique individuals have for building our lives, though not all options are equally represented (much less recommended). Regarding employable skills, in Venice we have the option of becoming glass blowers or gondoliers; in London, taxi drivers or museum guards; in Bar Harbor, tour guides, waiters, or fishermen. We shape our situated selves according to the opportunities and demands of the cultures we grow up in.

6-9. Calendar. Where clocks count out the hours, minutes, and seconds of our daily engagements, calendars sort out the days, weeks, months, and seasons of those engagements year after decade after century. No aspect of culture is more profound than the standards against which we measure our lives and endeavors, scheduling them in advance, checking them off as we pass hour-to-hour, day-to-day. Clocks and calendars provide standard changes for calibrating the activities of diverse people who need to synchronize their engagements in order to get along together more-or-less in harmony. Without them, culture would collapse in a melee of discordant wrangles, missed opportunities, and forlorn hopes. We would fall back on rough reckonings by the sun's position in the sky, comparative lengths of day and night, and relative warmth of air blowing against our skin.

Our parents duly check our first minute on Earth against clock and calendar, setting our birthdays forever after. I know the exact hour of my birth because the night shift of nurses at Faxton Memorial Hospital had left, the day shift not yet come on. The doctor handed a cone of cotton and bottle of ether to my father, telling him to give a whiff to my mother, which he dutifully did, missing the cone, pouring a stream into her eye—and out I popped punctually at 8:00 a.m.—a memorable moment for the few who were present. Here I sit at my computer eighty years later, my clock still ticking, calendar turning leaf to leaf, both now running faster than they used to.

Thoreau wrote in the concluding chapter of *Walden,* "If a man does not keep pace with his companions, perhaps it is because he hears a different drummer." All cultures are built on different ways of measuring time and days, so march to different drummers. The *Gregorian calendar* is based on a solar year of about 365.25 days, with an intercalary day added every four years. The *Islamic calendar* is based on lunar years of 354 or 355 days, so takes some thirty-three years to cycle through the seasons. The *Hebrew calendar* is essentially a lunar monthly calendar adjusted in keeping with a solar year by periodic insertion of intercalary months to keep religious festivals within designated seasons. *Hindu calendars* are not strictly lunar or solar, but are based on years of approximately 360 days having twenty-nine or thirty-day months, with one month being repeated in a given year now and then. The *Chinese calendar* is lunisolar, based on seasonal segments of approximately fifteen days. There are many other calendars in use, but the Gregorian calendar is increasingly relied on for international trade and transportation.

We tend to internalize whatever cultural calendar we are born to, so that we become calibrated by that system, which comes to seem the proper system for keeping track of birthdays, anniversaries, school years, vacations, significant holidays, sporting events, and so on. But calendars do more than that. They tie our inner workings to the workings of the solar system and the stars beyond, to the apparent motions of sun, moon, and planets, and to the seasons here on Earth, the migrations of animals, the fruiting of trees and flowers, the very possibility of agriculture. It took millennia of patient observation to understand how sunlight affects the Earth and every member of the human tribe. The purpose of most religious rituals is to synchronize human effort with the seasons, so securing us a place in the natural world that we may be fruitful and multiply.

Put simply, without some sort of calendar, most of us would not be here today playing computer games, surfing the Web, tweeting every notion that enters our heads—engaging our way through life as we do these days instead of stalking our next meal through the woods, dancing for rain, praying our seeds will come up, looking to the heavens with awe and dread.

6-10. The "God" sound lives on. Like almost all words, "God" is a word we are born to. A sound made by members of the community that receives us on particular occasions within specifiable situations. Our job, as novice members of our culture, is to figure out how to identify the situations within (and occasions upon) which it is deemed appropriate to make the "God" sound and gestures (closing eyes, looking up, folding hands, etc.). Those situations and occasions, then, in an experiential sense, are what "God" comes to signify or mean to us. When we utter "God" ourselves, we intend it to elicit the attitude we identify in ourselves on just such occasions.

If as children we are told, for instance, that God hears our prayers, knows our most intimate thoughts, and watches everything we do, then we come to fear "him" as a kind of extension of our most intimate selves from whom we can hide nothing because we are told he already knows all our secrets. That is what God comes to mean to us, the all-knowing authority who observes and judges us at every instant, so we'd better behave ourselves, particularly when our parents aren't on the scene. If we are told that God made us and everything on Earth, then he creates every situation we find ourselves in, and how we respond is a kind of test of how good we are in his eyes. As God

131

wills, we must humbly obey—such is the attitude we are to take toward the metaphor that we are told is our maker.

Good behavior, then is our culture's gift to us through the agency of the ancient God lore to which we are born. If we are told that all women bear the stain of original sin for Eve enticing Adam to know what must not be known, then little girls had better learn to be doubly good by scrubbing their minds squeaky clean to avoid playing the role of temptresses themselves. And so on. Simply by being born to the presence of such powerful words and ideas as God, sin, evil, lust, duty, and obedience, we become conscious of the meanings of those words in extremely intimate terms as if they had been in us from the start, as our heart, lungs, liver, and sexuality are givens at birth.

We are born to two worlds at once, natural and cultural, and become creatures of those worlds largely through our engagement with language as augmented by our unique existential experience. Much of human consciousness is spurred by comparison of our personal proclivities against the cultural narratives we are told as children, our struggles often reflecting attempts to narrow (or hide) the gap between them. It was the papal *doctrine of discovery* that authorized early European settlers to seize "New-World" lands held by "inferior" pagan tribes. A similar doctrine derived from prophetic teachings about covenants and heathens empowers today's zealously pious (backward-looking) Zionists to crush Palestinian survivors in the West Bank in the name of redemption of a Jewish land in Israel according to a wistful contract with their long-ago God. And the Islamic doctrine of jihad is twisted and updated to justify using commercial aircraft as weapons against the Twin Towers and Pentagon as divine punishment upon an infidel nation on September 11, 2001. God's vengeful spirit was perverted as the basis for our wholly unwarranted attack on Iraq in retribution for the felling of the Twin Towers—an event Iraq had nothing to do with. The upshot of our respective faiths being that our childhood fear of fictional gods makes terrorists of us all.

So do such engagements perpetuate discord between cultures on the basis of outmoded narratives of belief we instill in our hearts, and retell ourselves in hardening those hearts in order to do violence against members of other cultures. We do our best to keep past doctrines alive as guides to our ethical behavior in situations that did not exist when those doctrines were pronounced, but are justified on the basis of undying belief in great and terrible gods we claim as our special friends and protectors.

6-11. Primate see, primate do. In engaging with those in our immediate households, our young selves notice valenced discrepancies that shape what consciousness is able to do for us in seeking to become more like, or distinct from, those around us. *Mirror neurons* are brain cells that favor engagement. Apparently common in humans, they enable us to mimic (or perhaps inhibit) actions we see (or hear) others perform. Providing a direct link between perception and action, such neurons appear to jump-start engagement without our having to wait for our passions and meanings to mature. They are primate-see, primate-do neurons, which facilitate our initiation into human society. The bright eyes, smiles, laughter, coos, and hugs of our caregivers encourage us to respond in kind, telling us we are doing something right, so we happily adopt that as our practice because the spontaneous action and playful mood give us pleasure. We hunger for similar engagements because we thrive on participatory stimulation, preferring engagement to being left on our own. In musical families, we tend to take up the piano or flute; in professional families, we apply to med school or law school; in orthodox families, we adopt the prescriptive attitudes and beliefs of our parents and grandparents.

My father's mother died late on the day she gave birth to her son. The one person on Earth who was to have been there for him simply vanished. Others in the community rose to the challenge the best they could, but they all had prior commitments to attend to. It was not an encouraging situation (*see* Dedication).

The issue was survival. Others could do only so much. They pulled him (my father-to-be) through, but self-reliance was his take-home message, the one he modeled for his own family later on. When it was my turn to be born, my family resounded with the echoing silence of my father's earliest days. John B. Watson had recently (1928) published his *Psychological Care of Infant and Child*, which I will distill to the following paragraph:

> There is a sensible way of treating children. Treat them as though they were young adults. Dress them, bathe them with care and circumspection. Let your behavior always be objective and kindly firm. Never hug and kiss them, never let them sit in your lap. If you must, kiss them once on the forehead when they say good night. Shake hands with them in the morning. Give them a pat on the head if they have made an extraordinarily good job of a difficult task. Try it out. In a week's time you will find how easy it is to be perfectly objective with your child and at the same time kindly. You will be utterly ashamed of the mawkish, sentimental way you have been handling it (pages 81-82).

Disengagement between generations was the norm in those days, particularly in families susceptible to the lecturing voice of academia. That is the world I grew up in, so it is no wonder I sit here at my computer eighty years later, wrestling with notions of encouragement and engagement because I am so intimately acquainted with their absence through lack of emphasis in my formative experience. I was trained more as a rat than a person.

6-12. Humor. Jokes fire expectancy, hook us into engagement out of pure curiosity, and then surprise us when the punch line fulfills the promise in a way we had not imagined. The joke is in the sudden relief we feel when tension is released in a concise and satisfying manner that both meets yet exceeds our expectations. The self is initially situated within the limits and conventions posed by the setup, but then is yanked beyond those constraints by the neat resolution delivered by the punch line. Ah ha, I get it! Even though "it" has never been said explicitly yet is implicit in the space between the situation and its glib resolution. The joke, that is, is virtually contained in what has not been said, its absence being what makes it a joke.

> Fred is a fixture along this part of the coast. When you see a white boat just offshore, that's Fred, hauling lobsters. He's ageless, been around forever. So when he died, the editor of the local paper asked Myrtle, his wife, to write an obituary. She said she could do that, wrote two words on a piece of paper, and sent it in: "Fred died." The editor called her and asked her to expand on that. "Everybody knows Fred," she told him, "what's there to say?" "Some people might not know him as well as you do, Myrtle," he said, "flesh it out a little." So she did. The obituary ran, "Fred died. Boat for sale."

Then again, there's the pub called "The Silent Woman" which has a sign out front showing the body of a woman in long skirt and apron carrying her severed head before her on a tray. And Samuel Johnson's 1755 backhanded definition of *oats:* "A grain, which in England is generally given to horses, but in Scotland supports the people."

Every joke, like every metaphor, is a nonce theory of meaning; you have to be there (in that unique situation) to "get" or appreciate it. There's more going on than meets eye or ear: a full-fledged engagement teasingly situated in a past, present, and plausible future. Jokes and metaphors play spontaneously vis-à-vis our expectations on the leading edge of our engagements. They do the heavy lifting of revealing the future as an extension of, and variation on, prior expectancies. They help us thrust ourselves forward by being open to both past and future at the same time within the structure of the current situation as we construe it. Old jokes and metaphors are stale and uninteresting; what we crave in engagements is always the new, or at least new versions of worlds known by heart.

6-13. Music. Whether as listener or performer, nowhere are we more engaged both personally and culturally than in music. The rhythm punctuates the flow of our participation, beat after beat after beat, melody rising and falling as we go, harmony playing tag with dissonance, tone colors shifting through a palette of sounds made by saxophones, guitars, pianos, trumpets. In music, as in jokes and metaphors, the challenge is to attach the coming instant seamlessly to where we are now in our experience, and to where we have recently been. The flow is of the essence, like riding a wave or snowy slope. That wave is as much internal as external, for it is in us as we feel the full force of the drive and the lilt. Music is something we immerse ourselves in; it dissolves boundaries between people, between our innermost selves and our felt situations. It is not about anything but now, now, and now—the art of being engaged. It achieves its effect by being familiar and novel at the same time, heading us toward the future as an evolving possibility created by musicians themselves as they perform together, mutually conspiring in telling ways as they interact with one another on the spot, or as leader and composer would have them play themselves out. Music is body chemistry within and between members of an ensemble made audible in all of its nuances. Its power stirs us to realize our own body chemistry, getting us moving as we would move ourselves. Music, theme song of the life force, incites us to engage and be alive.

I write these words as a non-musician who is susceptible to small jazz ensembles and string quartets. I love the interaction between different voices as they move in and out of intimate relationships. The massed, brassy sounds of big bands turn me off like somebody shouting at me through a bullhorn. When as a kid I was sick enough to stay home from school, along with my ginger ale I got occasional doses of Gershwin's *Rhapsody in Blue* and Ferdie Grofé's *Grand Canyon Suite* (I still hear those clip-clopping cocoanut shells) via a radio station in Syracuse, which I referred to as "dirty music" because I didn't much like it, so chose silence instead. But the prevalence of music around the

Earth tells me that rhythm, melody, harmony, and tone color contribute to a language of sound inherent in consciousness itself. We are born to music in the guise of howling winds, pelting drops, bird calls, trumpeting elephants, pounding hooves, lullabies, love songs, and dirges. If these didn't exist, we would have to invent something like them. Which we have, over and over again in every highland and lowland on Earth.

Our sensory-motor engagement with music is scarcely believable. Eyes, ears, lungs, vocal cords, fingers, feet—that's our basic equipment—shared by Bach, Beethoven, Berlioz, Ravi Shankar, Woody Guthrie, Billie Holiday, Stevie Wonder, and every other musician and would-be musician through the ages. How they make the world sing! And dance. And clap its hands. Us along with the rest. When I think of the concentration it takes to develop the skill of playing a musical instrument, it shocks me that even one person would work that hard—yet billions make and consume music every day. Making music is the opposite of making war. Constructive, connecting, civilizing, life-affirming; not destructive, shattering, dehumanizing. When millions of CDs and records are sold, think of the lives enriched, perhaps changed forever. As I was changed one day in 1952 when I walked off Massachusetts Avenue into an open rehearsal of the Boston Symphony Orchestra under Pierre Monteux playing Berlioz's *Symphonie Fantastique.* I sat in the back row of the orchestra seats and could hardly believe my luck. My brain was rewired on the spot. That was sixty years ago; I've never forgotten the sights and sounds I stumbled onto. I am born to music as I am born to rhythmically swing my legs or utter one syllable after another.

People make music; music makes people. However it works, music is within us. Rhythm, melody, harmony, tone color are inherent in our being conscious. Think music, poetry, dance. Think smiles. Think passion and meaning combined. Think applause. Think engagement.

6-14. Haiku I. Poet Matsuo Basho refined haiku as a language for jotting down notes about his adventures traveling through seventeenth-century Japan. He combined his cultural understanding, language, and sensory impressions to be sure of capturing all three in concise, evocative, and memorable form. Millions of haiku (and formulaic imitations) are written every day, testifying to the success of his method. Like music, metaphor, and humor, haiku is a medium relying on qualities of the mind itself as a basis for engaging with others on intimate, experiential terms. In Japan this has become a cultural means of engagement that does not readily translate into other languages.

As Basho (1644-1694) wrote them, haiku were a forerunner of tweets in being a means of capturing impressions while away from home, getting just enough down on paper to kindle memories that could later be shared with others. It's tricky translating haiku from the Japanese, not only because the syntax differs from English (verbs coming at the end), but there are no articles (a, an, the), pronouns (he, she, they), distinctions between singular and plural, or marks of punctuation. Which often get added in translation to make them acceptable to those used to hearing and writing English. Instead of punctuation marks, haiku include *kireji,* words that serve as a kind of stage direction regarding emphasis and rhythm to be read aloud without adding meaning to the text. Because of such inherent differences between Japanese and English, I ignore the formulaic 5-7-5 syllabic structure, but retain the customary use of words evoking natural scenes, particular seasons, or times of day.

I think of haiku as the ultimate distillations of consciousness. Grappling with *becoming aware of being aware*—the so-called haiku moment—haiku are meant to capture what it is about a scene that attracts our attention and draws us out of our everyday selves, heightening our engagement with life. Our sensory impressions, everyday conceptions, understanding, feelings, and spectrum of personal values are all brought to bear in reading and writing haiku. They address the exact moment we become alive to ourselves in rousing from our streaming reflections to discover we are participating in a situation of particular note. It is a haiku's challenge to capture that situation in the most telling way possible. Working from Harold G. Henderson's Japanese-English transliterations (*An Introduction to Haiku*, 1958), I am not able to be true to either Basho's art or his experience. I offer rough paraphrases instead, my naïve renditions of what I imagine travels in ancient Japan to have been like.

> as for the blossom
> by the roadside
> my horse ate it

The wildflower by the side of the road attracted both Basho and his horse's attention, leading to, first, the horse eating the flower, and then Basho capturing the incident in (the original of) this haiku. The attractiveness of the flower was a setup for its demise, producing the surprise and irony that made the incident stand out in Basho's mind. This is precisely the kind of moment that wakes us up because of the disparity between appreciating one of nature's beauties and then witnessing its inglorious fate. Compressed into a single episode of consciousness, we immediately grasp the universality of blooming and succumbing.

> hollyhocks steer
> by a hidden sun
> spring rain

Basho here draws attention to the sun's location in the sky and the direction hollyhocks face in tracking it—even though it may not be evident to those who do not depend on photosynthesis to make their own food. In this case, rainclouds hide the sun, but the hollyhocks spy it out and face toward it nonetheless. As hollyhocks turn to the sun, Basho turns toward the hollyhocks. We all have our tropisms, deliberately turning to face that which attracts us. Our loops of engagement echo that natural force, ensuring we seek the force that sustain us—food, air, water, companions, shelter, children, health, safety, beauty, and other drives and values that direct us toward what we need to survive.

Hollyhocks need sunlight, Basho needs hollyhocks, we all seek engagement with what keeps us going.

> windy night
> Sado Island awash
> in the Milky Way

Here Sado off the northwest coast of Japan provides an earthly reference point for Basho's otherworldly apparition. It is night. Wind is blowing. Seas are heaving, crashing. Over Sado, stars gleam in a swath across the sky. A night to remember. So Basho jots down a few words to spark his memory later on when he feels moved to recount his adventures. His life is one momentous journey made up of experiences such as this. Imagine what it was like in those days before the advent of radio, films, TV, computers, and the Internet—the endless stream of distractions via new media meant to capture our attention. Basho's words flow from a world different from ours of today. But his jottings are still with us, and we can recover some of the world he knew directly if we will apply ourselves to that task.

> where cuckoo flew
> a far island
> looms

Here the cuckoo flying into the distance leads the poet's eye to an island he has not noticed before—which might be where the bird was headed in the first place. We talk about what William James called the stream of consciousness—as if awareness flows by itself. But in truth, we are responsible for the sequence in which we become aware of events because that sequence reveals how we direct our attention from one salient event to the next. Events don't flow; we flow. Consciousness streams within us as we are moved to track the changes we notice. We are made to discover motion in our surroundings—such as the fly we spot out of the corner of our eye, birds winging into the distance, islands rising from the sea in that direction—as I once discovered the full moon in the exact spot where a passing hummingbird vanished. Replacing a near bird with a far island in our attention is no mean feat, yet we perform similar tricks many times a day. One thing points to another, and that to another. Think of movies, television, videos, auroras borealis. We are hooked on motion and tracking change, which we interpret as plots and narratives, and sometimes haiku.

> farmer's child
> poised to husk rice
> looks to the moon

A depiction of Basho's experience at the time, this highly compressed and economical image of a boy working in a field was seen as "poverty's child" (*shizu-no*), which I see as a melo-dramatic stretch, so cut back to "farmer's child." The situation drawn to our attention is that the boy turns from his work to gaze at the moon, betraying a conflict of attention that rouses the poet to take notice of the boy's double-edged stance of reluc-tance and yearning at the same time. That feeling is not stated in so many words, but is conveyed by the twofold action at the heart of the poem. Multitasking bears a message of its own.

> dry branch
> crow settles down
> autumn nightfall

The power of this poem flows from its sound in Japanese, necessarily sacrificed in translation: *kare-eda* | *ni* | *karasu-no* | *tomari-keri* | *aki-no-kure* (withered-branch | on | crow's | settling-*keri* | autumn nightfall). The image is all of a piece, the coming of night paralleling and affirming the crow settling on its branch, and vice versa. Crow is to branch as night is to autumn; or else branch is to autumn as crow is to night. Either way, the imagery is tight knit. It is the coherence of both words and image that heightens our experience.

> rural Japan
> art flows from
> rice planting songs

A song Basho heard in the northern province of *Oku* on one of his journeys is at the heart of this haiku. He composed the verse on the spot when asked to share a poem he had written in the northern interior. He heard in the local work song the deep beginnings of the more sophisticated art of the city. The first part of the poem stems from his understanding of the customary transition from rural folk art to a more intricate urban treatment. He contrasts the two—song and art, country and city—and that contrast creates the tension between the two parts of the poem that whets our attention. The first part is conceptual—art has to have begun somewhere. The second is based on personal experience during his journey. The result is a larger understanding of the spatial and temporal relationship between rural and citified Japan.

> summer weeds
> fallen warriors'
> relict dreams

On one of his journeys, Basho came to the site of a famous medieval battle at a former castle, in his day gone to wasteland. He wept at the thought of lost glories and wrote this poem on the spot. Vegetation was all that was left of the famous story, providing the basis of the concrete sensory impression that contrasted with details his imaginative understanding recalled from memory. Descended from warriors himself, Basho identified with those fallen in battle, and felt their defeat personally. That contrast stirred his consciousness, and he compressed the disparity between then and now into this poem.

My point here is that the world we truly live in is in the felt depths of our consciousness. When we die, all trace of that lived consciousness disappears, leaving our families, friends and acquaintances to carry on as they will. It is highly unlikely that anyone would be able to recreate our unique awareness from the works, notes, and scraps we leave behind, for no collection of physical objects can capture the essence of a living being. But if we read between the lines that a singular mind has left us, we can attempt to construe what it was like to live that particular life. I am grateful to Matsuo Basho for suggesting what might be captured through supreme skill, effort, and economy. I see his legacy as a gift to those who dare to accept it.

Basho engaged the landscape of his century through his travels and teaching. His loop of engagement was extremely active and discerning to the end. He used horseback technology, language, and little else in leaving us a portrait of one man's sensibility and daily engagements.

6-15. Haiku II. I know way more about the following haiku than I do about Basho's because I wrote them in English, my native language. I have not just the words but recollections of the scenes and feelings that spurred them as well. I offer them here as nodes or nuggets of my lifelong engagement with situations I find myself in. I do not mean to compare myself to Matsuo Basho, more to shed light on the makeup of engagements them-selves, which sometimes open us to poetic truth.

> autumn dusk
> contrails of cargo plane
> back from Iraq

The great circle route from Europe crosses the Maine coast where I live, so contrails are common when air traffic is heavy. In this haiku, sunset and contrails were announced by sensory patterns as I walked through a field at dusk. I could make out the silvery plane, but couldn't tell if it carried cargo or pass-engers. My mind, however, was steeped in the Iraq war, an echo of my situation at the time, which I projected onto the contrails. I imagined bodies of fallen soldiers being ferried home for burial. The words came to me in a flash as my truth at that moment of experience.

> red year
> berries, maple leaves
> blood of war

Again, the situation here is dominated by the war in Iraq. To me, the world was blood red. The berries and leaves were sensory impressions; the blood flowed from my situated self, coloring the entire year as viewed in the fall. These lines could apply to any year in the first decade of the new century.

> old stump
> fall shadow
> cut short

Why write about an old stump? To me, stumps are lives cut short, shades of what might have been. From the trail I was on, seeing a stump lit by a low sun, I immediately missed the shadow it would have cast had that tree been let to stand tall. Haiku are often about the blend of concrete sensory impressions and their effect on our imaginations. In this case, I was engaged with—and spoke to—the cutting of that tree by way of its truncated shadow. As Thoreau wrote in *The Maine Woods:* "A pine cut down, a dead pine, is no more a pine than a dead human carcass is a man."

> five yellow leaves
> fall in the road
> the road vanishes

Walking along a road through yellow woods, I watched five maple leaves waft into the road ahead of me. When I reached them, all I saw was the constellation they formed, the road on which I traveled having no more substance than the dark sky at night. This is the familiar figure-ground effect that occurs in our minds as we strive to make sense of what we "see" before us.

> autumn now
> keys to forgotten doors
> clink as I walk

Yes, it's true, the older I get, the less I remember why I do what I do. Autumn announces the aging of the year, and of my very mind and bones. I carry keys to open important doors: home, office, files, padlocks. Those keys tell me who I am in terms of the places they give me access to. Rather, who I once was. My keychain now is largely symbolic. I lug it around out of habit and sentiment, clinking keys taunting me at every step, telling me I'm not the man I used to be.

> November
> all vanes point
> north

It's all north's fault, cries every weather vane in town, pointing to Arostook County, Canada, and the Arctic beyond as the wind rises and temperature drops. That's how I take it. The vanes are merely doing their duty of pointing out the direction from which the wind blows. My situation is that I'm cold and want something to blame as the cause of my discomfort. I am the accuser, not the vanes. In the haiku, I don't even mention the wind.

> blowing snow
> no place now
> for wind to hide

The synchronized motion of blowing flakes is a salient feature of sweeping or swirling snow, making the wind visible in the process. I am as fascinated by blowing snow as I am by rolling swells, crashing waves, rushing tides, flights of birds—all visible signs of motivating forces that drive life ahead, ever ahead. Or sideways. Or curled into itself.

> at five below
> no apple on the bough
> purple finch

On my first glimpse through the window in the morning I was stunned to see a purple finch fluffed into a ball to keep warm, and not one red apple on the tree. Clear as day, winter was here. I love the finch for its singing, but, too, for its stamina. It is a survivor, adapted to the prevailing climate of Maine. I identify with that—with *being here* where we live, finch and I, as expressions of our place on Earth.

old table
friends sit and talk
grain of the wood

I took a bus to Blacksburg, Virginia, to visit Jean and Gene Franck, friends I had not seen in over thirty years. We ate at the same table they'd had on Tufts Street in Cambridge when Gene and I worked together. Same table, same grain, same friends—it all came together, the past as remembered amidst the immediate, sensory present.

summer pond
striders dart
cloud to cloud

One August, members of the haiku group meeting monthly in Bangor Public Library agreed to make an excursion to gardens in Northeast Harbor for the purpose of writing and sharing haiku. Where Basho was moved by the sound of a frog leaping into his famous "old pond," I was fascinated by the lilting spurts of water striders thrusting themselves at the center of their wavelet halos about a pond in the Azalea Garden where we stopped. The geography of the pond was laid out in terms of reflections of overhead clouds in a blue sky. From my situated perspective, clouds and striders resided in a single pondscape in my mind.

I offer these several poems here to demonstrate that haiku moments occur not in the world but in human minds resounding with cultural and personal overtones. These combine with sensory phenomena to present small, meaningful worlds nested in felt situations. Each haiku is a window of words opening onto such a world in a particular mind, as I described an apple-sized purple finch fluffed up on the branch of an apple tree one cold winter morning in Maine. Just as every haiku is personal, every engagement in every medium is personal as well. We put the stamp of our personality on our experience, and call it a world.

Every haiku, every engagement centers on our *being there* holding our own in such a hybrid world for one telling moment.

6-16. The news. The workings of our minds—what we are engaged with—make front page news in every journal in the land. Our minds are told in so many column-inches in full public view: obituaries, fires, explosions, chemical spills, accidents, shootings, knifings, mass killings, foreclosures, bankruptcies, layoffs, graft, theft, collusion, domestic violence, child abuse, greed, jealousy, retribution—laid out for us to peruse at our leisure. But it's not all catastrophic. There's a good side, too: promotions, hirings, sports, new construction, awards, improvements, talks, entertainment, art, graduations, appointments, victories, medical advances, humor, success stories, engagements, weddings, births, discoveries, rescues, faithful pets, news of our neighbors—again, all there to keep us informed of the good side of life.

News, in short, triggers consciousness by drawing attention to what's worse than usual or better than usual. The usual itself is the forbidden middle; nobody cares about that, so we take it for granted. What we want to know is what's exciting and what's terrible. A workman-like performance is the kiss of death; we want catastrophe, we want ecstasy, we want superlatives! Why is that? Because we are built to be conscious, and extremes demand consciousness if we are to deal with them effectively. Which, after all, is why we are here—because we've gotten this far by coping every week with whatever comes up and aren't about to quit now.

Business, sports, the arts, politics, education—we engage in enterprises that require skill and great effort just to stay in the running. Nobody works harder than an athlete—unless it's an entrepreneur, dancer, politician, teacher. Or mother raising kids for that matter. Or father trying to make ends meet. Or student in elementary school, middle school, high school, college, grad school. We survive by our wits, and our wits are fueled by consciousness. Stupor and oblivion stop us in our tracks.

6-17. Nature. If culture influences our engagements, nature does so even more profoundly. Food, air, water—we engage with them or die. As a photographer, I am primarily engaged with the natural world of coastal Maine. With every muted click of my shutter, I celebrate the land-air-water where I live. I don't press my finger unless I am moved in a meaningful way. And what moves me is my witnessing the natural world of plants, birds, animals, insects, estuaries, forests, watersheds, habitats, weather, seasons—the world that provides me not only with food, air, and water, but passion and meaning in my life.

The word religion stems from the Latin verb *religare*, meaning to tie fast or tie back. Tie to what? Community, culture, ardent belief. Its undertone is to belong as a member of something larger than yourself. Nature is my religion. My particular culture is arbitrary, a function of where and when I was born. But nature's hold on my life is absolute. I am alive among other people, creatures, and vegetation by the grace of nature. Nature gives me being, existence itself. Religion might modify that existence, but it stops short of making my life possible in the first place, and of sustaining me as I grow. Nature is the great mystery. Religion is a tale told by high priests and prophets.

Nature as we know it includes a range of levels from sub-atomic particles, atoms, molecules, via organic and inorganic routes up to our planet, our solar system, our stellar neighborhood, our galaxy, our galactic group, on up to the entire universe, and beyond—if there is a beyond. I live largely in the company of organic beings such as clams, crabs, ants, butterflies, fish, frogs, crows, dogs, people, grass, trees, in habitat areas such as mountains, valleys, streams, rivers, lakes, shorelands, estuaries, bays, oceans, and continents. These are a few of my haunts and acquaintances. I don't know them as they exist in themselves; I know them as impressions I am aware of courtesy of my sensory receptors, which translate energy given off or reflected into the language of qualities that tell me something is out there on the far side of my senses. What an array of such qualities might possibly *be* or *mean* in my personal experience is up to me to decide in light of recollections from earlier days.

I am kinetic, everchanging, and nature is, ditto, so my consciousness is the intersection where my changing mind intercepts perceptual signs of a changing world through the host of engagements I am privileged to entertain in the course of my life. Nature and I go together, accompanied by the cultural setting I am in at the time. It is difficult to say exactly where the self that I am, my mind, my culture, and the natural world separately begin or end. I am an aspect of my mind, culture, and nature, and they are respectively aspects of me. We complement one another, but, too, are engaged as parts of one another. Like Ping-Pong balls being tumbled in a whirling wire cage, it is hard to say where self-mind-culture-nature pick up or leave off. Consciousness embraces all four, and I know only that it does, not how it might accomplish such a miracle. This book is as close an account as I will be able to give in this life.

6-18. Age. Growing old can have two opposite effects on our wits, either sharpening or dulling them. In the second case, a person can pale to a hollow specter before our eyes. I watched Alzheimer's steal my mother from this world over a period of five years. In the end when I'd visit her, she'd ask, "Do I know you?" Her engagement with life did a slow fade until she had nothing left and was as good as gone. Then, poof, she was gone.

I seem to be exhibiting the opposite effect in that what wits I have left are keener than before, perhaps because many of my trivial concerns have fallen away, leaving the core issue, *consciousness itself*, naked, raw, exposed to my self-reflective temperament. If each day dawns on a smaller view than the day before, that view grows sharper and clearer. I am beginning to see the light. At least *my* lights—whatever it is that illuminates the workings of my own mind.

I am still able to recall once-familiar names, and am not aware of growing gaps in my memory. Things come to mind as they always have, appearing through the gate to whatever situation focuses my thoughts and attention. As ever, my thoughts feel connected, focused, and germane to my interests. My operative assumption is that I am not losing my wits with age, or at least not yet. So I'll stick to my belief in the coherence of my thoughts. Which gives me courage to keep writing these words based on the study of my own mind and its unique stream of consciousness. And, beyond that, to recommend introspection to others who may feel led to do the same within their respective fields of awareness.

I am searching for the motive force behind our loops of engagement. In experiential—not scientific, philosophical, or psychological—terms. To me, perception, situations, and actions are all of a piece, a dynamic progression channeling a coherent flow of energy. But am I correct in maintaining that the brain is a comparator drawing attention to disparities between our expectancies and sensory impressions? And further, that the mind is driven by the results of such comparisons round after round in just a few seconds? I get to the point of sensing a cohesiveness in the overall process, which gives me hope, but I can't pin down its origin in neurological terms because I am no neuroscientist.

Here I will simply note the growing sense of cohesiveness I discover in my own mind as I age, and leave it at that. My thoughts seem to resonate with one another—sensations, situations, meanings, actions, engagements—as if different modules or nuclei in my brain were working cooperatively together, adding to a larger whole than they would by operating independently or in opposition. Experientially, I take that resonance as an indication I am onto something, not as an illusion left in the wake of a weakening mind.

6-19. Aesthetics II. My approach here is experiential, not philosophical, psychological, neurological. But if I had to choose a discipline from among ontology, epistemology, ethics, aesthetics, logic, sociology, and all the rest, I would favor aesthetics because my attention is so drawn, not to beauty, but to relationships. I am interested in why things hang together—pictures, melodies,

flavors, memories, social groups, ideas, my streaming consciousness, and so on. Here it is again, cohesiveness. If every round of
engagement were unlike every other, experience would be cha-
otic—but it's not. Experience, at base, is meaningful in forging
links (connections, relationships, harmonies) with that which has
come before. I view those links not as logical connections but as
fitting into a pattern of relationship at a particular level of detail.
Connections. I'm into making connections, appreciating connections, understanding connections. Which to me is the business of
aesthetics. Of inherent affinities. Mutual attraction and affirma-
tion. And of inherent distinctions. Mutual separation and integrity. Yes, based on similarities and differences. That is, on comparison, on what I view as the spring of my own streaming consciousness.

6-20. Life and death. When engagements are blocked—as when
the NYPD forcefully descended on the occupiers of Zuccotti
Park—feelings run high. People don't like to be interrupted,
much less stymied or taken to jail. We are born to engage.
Engagement is life; every setback an intimation of death. That's
why thwarted engagements become social causes. Think civil
rights, gay and lesbian rights, women's lib, abortion rights,
immigration policy, the Arab spring, even losing streaks suffered
by local teams. From a personal viewpoint, our most fundamental freedom is the freedom to engage as we see fit. Socially, our
basic principle is to engage others as we would have them
engage us. That is, to level the playing field, return the favor,
reciprocate; play fair. Violence upsets the balance, turning social
engagements into conflicts that can easily get out of hand. In a
communal sense, engagements are projects in working *with* one
another for mutual benefit. In conflicts, we work *against* one
another, aiming to belittle those we do not identify with, to
dominate them as losers under our sway.

6-21. Love. Love is an invitation to discover deeper levels of
engagement than were thought possible, leading to a trust in,
and commitment to, experience on those levels. The situation

opens up, bringing a new sense of resonance between all elements of consciousness. Energy is released. Doors open upon exciting new ways of being ourselves. We become new persons, released from bondage to the familiar routines of the past.

Frustration, on the other hand, is engagement precluded, leaving the self locked in despair. Anticipated possibilities are shut down, leading to losses of trust and commitment. Dissonance becomes the primary quality of experience. Energy dissipates. Doors close upon hoped-for ways of self-realization. We feel old and worn out, condemned to a life of loss and disappointment. Our world has crashed; we sing the blues.

6-22. Power and wealth. It shocks me to realize how obsessed our nation has become with politics and economics—with issues of power and wealth. I do not see myself as primarily concerned with these matters, but the magazines I subscribe to treat me as if I should cultivate such concerns. I vote, pay my bills, do my work, and love my family. That, I feel, should be the end of it. But it's only the beginning. As more and more people can be reached at less cost and effort, the rich and powerful are increasingly dominating Earth and its people, treating them as subjects of a single kingdom, ruled by themselves at the top—much as Slobodan Milosevic ruled Serbia, Muammar Gaddafi ruled Libya, Hosni Mubarak ruled Egypt, and the likes of Yugoslavia's Tito and Romania's Ceausescu ruled in Eastern Europe.

A dim view? I have encountered this tyrannical attitude already in my dealings with Microsoft, Apple, Google, Adobe, Facebook, Wikipedia, and other shapers of the Internet. Every time I send an email, I am treated to an advertisement triggered by a secret scan of my message. If I "like" something on Facebook, that becomes data to be sold to the highest bidder. Every purchase I pay for with my debit card tells interested parties what I paid—for, where, when, and how much. Who am I? Just one more pocket to be picked. Surveillance is the future of the Internet. We engage electronically at our risk. Is there such a thing as public privacy? In the cloud, it's already here.

The recent recession is a conspicuous example of power and wealth run amok, revealing collusion between banks, rating agencies, insurance companies, business schools, corporations, lobbyists, and their respective regulators. The people suffering most are those enticed to invest their money, pensions, homes, and labor in the system, as they are invited to do by all those happy, colorful advertisements and brochures. Even U.S. Presidents tell us to support the national economy as our patriotic duty, even as corporations ship the jobs we once held overseas, and shirk on paying their taxes. As a people, we are paying more, earning less—or not working at all. As our population grows, we will increasingly be tiered in a top-down structure where might and power accrue upward in exchange for weary spirits and bodies in the depths. We may have outlawed slavery in our society, but increasingly the rich and powerful are treating the people as their private property, which is a slaveholder's view of those whose every engagement they control for personal benefit.

Yet here I am claiming that consciousness is a matter of individual self-governance. What do We the People have to fear from governments and corporations? Getting fleeced alive, no less. I read that David H. and Charles G. Koch of Koch Industries each netted over seven billion dollars in 2012. What did you make in 2012? Are the Koch brothers a more worthy form of humanity? If governments and corporations care about people, it is the people they are trying to buy, or already own. More and more, our personal value is gauged from above—through the eyes of those who profit from owning our souls and allegiance, the very people who are blind to the personal characteristics, virtues, and vulnerabilities of humanity in the fullness of its diversity. The Supreme Court can no longer distinguish between mortal beings and corporate entities, or between money and free speech. Looking out from the highest bench in the land, the Justices get the impression that, since people and corporations all look alike, they deserve equal treatment before the law.

After building a case for engagement, I have to admit that there are two sorts of people (or two modes of being), those who can actively engage with others, and those who know only how to assert their own views without considering any response they might get. Perhaps this is a matter of hormones, estrogen promoting engagement, testosterone pushing assertiveness. Since we all have some of each hormone, our ability to engage may come down to the relative balance between the two in our particular instance. How would assertive types such as the Koch brothers, Mitt Romney, and Slobodan Milosevic measure up against listeners like Obama, Oprah Winfrey, and Bill Moyers? Do they even live in the same world?

I visualize a personality profile based on four basic types of engagement: *playing, working, caring,* and *asserting.* It probably already exists. I predict members of the power elite would stack very distinctly against the caregivers of the world.

The thrust of my remaining life energies is to get people to know themselves as individuals, to take responsibility for their actions, and to be wholly who they are in dealing with the combined forces out to turn them into consumers, voters, rent payers, box holders, residents, and other anonymous, weak, and impoverished examples of humanity in the mass. If our singularity is in our genes, bodies, and our minds, let us claim uniqueness and diversity as our defining strengths in opposition to those who reduce us to uniform standard products to be treated all the same. If we do not protect our right to engage as unique persons, we might as well not exist. Being equally free does not make us all the same. We are a people, not a mass market—a congeries of distinct individuals. Our situation is that simple, and that scary.

6-23. Creativity. Our engagements are told by what we do in the world. We alone know how we come to those actions. If we depend on others to tell us what to do (to program us), we are not being our own unique selves but are closer to robots, automatons, or droids. Our distinction is in our uniqueness, originality, imagination, and creativity—in a word, in our ingenuity—not our duty according to someone else's plan. Discovery is earned through exploration, and since change is the one constant in the universe, every day is given to us as another chance to explore—to engage—where we are, as who we are. If we're stuck in the same place, or as the same person, we are facing backward, not forward, going nowhere. What kind of journey is that?

Our job is to be who we are. That is, to engage as only we can engage. Yesterday, today, tomorrow. Again, again, again. There you have it, a life lived as yourself, not anyone else. You fit my definition of creativity by making and occupying your unique niche in the universe. No galaxies are alike, no stars are alike, no planets, no continents, no creatures—including you and me. Our roles are to play ourselves, not some character we pretend to be for the sake of its effect on others. There is no script we can follow to live happily ever after. Life is a voyage of self-discovery, not assigned labor. We find out who we are by living the life that we do, the life we imagine for ourselves by our own lights, and then enact in the world.

From the jottings of Matsuo Basho to the collected works of William Shakespeare, all of literature is a form of consciousness studies giving glimpses into creative human minds. Hamlet (to choose a single example of Shakespearean creativity) is portrayed as indecisive, unable to consummate his intentions as done deeds. In Act I, the ghost of his dead father gets him to commit himself to avenging his father's foul and most unnatural murder by Hamlet's uncle, out of lust for Gertrude, Hamlet's mother. Hamlet vows to take action, but swears his companions to secrecy as if the ghost—and the murder itself—had never happened. His hoisting the burden of his filial duty on his own shoulders winds the tension that drives the play ahead, priming our expectation to see the job done. But Hamlet keeps holding back, in Act III voicing his doubts in the famous "To be, or not to be" soliloquy, which reveals the worm of reluctance working within him, tempting him to suicide rather than face the consequences of his vow to his father's ghost. How is he going to do it?, we want to know. But he doesn't do it; he dissembles, and strikes all about him, including his sweetheart Ophelia, as coldly preoccupied. By now we are inside Hamlet's head where the action is taking place—if it is to take place at all. The mission is clear; the will to perform it has gone missing. This is not drama between separate characters so much as psychodrama within a fraught situation in the embodied mind of one man. The cruel resolution of that situation comes in Act V, the result of hesitation and conflictive subtleties leading to an inadvertent massacre of those assembled, good and bad alike, the upshot of Hamlet's reluctance to engage. Yes, his father's murder has been avenged, but at high cost to the innocent—Ophelia, Polonius, Laertes, Gertrude, and Hamlet himself, who fumbled the force of his own passions. All enacted by Shakespeare's situated self in the depths of his creative imagination. I see Aaron Swartz's death as *Hamlet*, Act VI, tale of yet another conflicted soul.

As a nonfictional character, author of himself, Steve of Planet Earth is a noticer. He pays equal attention to what's going on around and within him. His life is a series of sensory patterns, situations, meanings, and actions. The flow of energy through his mind makes him who he is, a potential engagement on legs, up and looking around for something of interest, something to care about and respond to. Since that flow is unique to him, he can't help but lead a creative life by being the noticer he is, putting his few skills into action, finding meaning all around him. He photographs wildlife and other natural subjects, makes presentations, writes up his discoveries, and is fully himself at both inner and outer poles of his personal awareness. He doesn't *follow* a particular path so much as *discover* it as disclosed by the life force within. He keeps track of who he is, where he is, in order *to be there* as himself, leading, he feels, a vivid, exciting, and productive life in the process. A life of passion, meaning, imagination, creation, and above all, engagement. ◯

Chapter Seven

REALITY

7-1. Levels of reality. The reality we find depends on where we look for it. In our minds there are many realities spread between sub-atomic particles and the grand expanse of the universe. Our word "atom" stems from Greek *atomos,* meaning indivisible, but atoms themselves turn out to be handily divisible into electrons, protons, and neutrons, and the latter two into ever more minute particles known as hadrons, all the way down to hypothetical strings, which may or may not be indivisible.

In the organic world, too, there are many levels of reality. Base-pair combinations in our genomes govern the ordered assembly of amino acids into proteins, which give structure and function to every cell in our bodies and the biological world beyond. Cells add up to organs, organs connect in organ systems, systems cooperate to form organisms, organisms group into families and societies, societies combine to form populations. Other planets doubtless have populations of their own, probably not much like those on Earth because conditions would be different, but perhaps based on similar chemistry. We'll probably never know for sure. The point about reality being that what you find depends on where you look, and that depends on the curiosity behind the looking, so for starters I'll say there are likely to be many subjective realities, not just one that serves for all.

7-2. Perspectives. It takes a situated observer with a particular outlook to speculate on the nature of reality. From an experiential point of view, without observers, there would be no reality in a mindless universe. If we assume there might be a reality out there nonetheless, that assumption would be moot because, if we didn't exist, we wouldn't be around to argue that or any other assumption. Neglecting the presence of subjective observers only leads to incorporating unwarranted assumptions into the nature of reality which, then, couldn't be real in-and-of-itself apart from such observers. But as I maintain, each and every human (since such do exist) speculates from a situated perspective based on a repertory of unique experiences. Therefore realities must of necessity be perspectival, requiring the outlooks by which they are viewed be made public for the sake of transparency and meaningful discussion or replication.

Reality is not in the universe, it is a labeled concept in individual human minds, likely to be different in each case. God is another labeled concept, so doesn't count as an independent observer free of situated thought and experience. There's no getting around the fact that thoughtful observers have a history of earlier thoughts and observations, all of a subjective nature known (if known at all) only within particular streams of consciousness. Which clears the way for there being a multiplicity of different realities, and blocks the way to there being any one general or universal reality as the product of a clumping process by which individual things are grouped (thought of together, conceptualized) to reduce or obscure variation between diverse entities. Talk of the *All* or *oneness* of the universe is largely empty in precluding discussion of the very details that capture our notice, including our minds, situations, and repertories of intimate experience.

7-3. Schools of reality. There cannot be as many realities as there are things perceived in human minds. If there were, that would eliminate reality as a meaningful concept since there could be no possibility of shared understanding between different minds. But given the subjective nature of consciousness, neither can there be a single universal reality that unites the perspectival experience of all people regardless of age, background, and outlook. Is there a third way between the singularity of individual experience and universality of common belief? As an exercise, I ask you to bear with me here as I list eight possible takes on reality, including the two extremes I have mentioned. Here is my menu of plausible realities:

A. Ideas, concepts, and universals (existing only in minds) as the basis of so-called *realism*.
B. Individual things as named (separate from things as perceived) as the basis of so-called *nominalism*.
C. Named concepts or groups of things in the world as a conceptual variant of so-called *nominalism*.
D. Proper names of individual entities as a kind of *proper nominalism* recognizing the uniqueness of at least some things in themselves.
E. Unique sensory patterns or impressions in the mind as the basis of perspectival *phenomenology*.
F. Categories or concepts as groups of common sensory patterns in the mind as the basis of a *conceptual branch of phenomenology*.
G. Felt and meaningful experience derived from sensory impressions as a kind of *experientialism*.
H. Practical or operational implications of sensory experience as a mindful kind of *pragmatism* or *consequentialism*.

These eight different kinds of reality are of three types: 1) sorted, conceptualized, or categorized realities in the minds of human observers (A., F.); 2) particular sensory or experiential impressions in the minds of individual humans apart from how they may be conceptualized or categorized (E., G., H.); and 3) grouped or unique entities identified and given names by humans (as if naming bestowed reality) (B., C., D.). The difference between these three comes down to the question of how reality is to be determined:

o by naming
o by categorizing
o by perceiving
o by function, meaning, and/or significance.

The concept of reality has a checkered past in being viewed in many different ways by different people in different places at different times throughout history. I acknowledge how confusing these considerations are in comparing machine bolts to chestnuts, and will decide the matter much as Alexander cut the Gordian knot by declaring that G. or H. (*above*) are the only options that make sense from my phenomenological perspective, and that I personally favor G. as the only choice I find consistent with my emphasis on the wide range of possible engagements by human beings in their respective situations. To each of us, *the real is that with which we are engaged at the time.*

The fact of engagement with an unknown other bestows reality as a quality of experience.

I give that sentence a paragraph of its own because I believe that whatever we pay attention to is operationally real to us at the time. Conversely, what we do not attend is unreal at the moment, but may become real through redirection of our attention. Reality, that is, shifts moment-to-moment as we strive to keep up with our streaming thoughts, actions, and consciousness. Alexander wanted to rule Asia; he had been advised that he who undid the Gordian knot would be the next ruler; he drew his sword and cut the knot. The rest is history. Real to Alexander, real to us. Not that reality in the mind must play as reality in the world. There is a disconnect between mind and world, but wisdom gained through experience helps us learn to bridge such gaps.

7-4. Bioenergetic engagements. Our engagements entail both an intake and output of energy, so in an operational sense, energy flows through our minds in deciding a course of action appropriate to whatever situation we gauge ourselves to be in. We interpret incoming energy in light of our experience of similar patterns, construe a situation from that pattern, understand what that particular flow of energy means, and decide what to do on the basis of our history of life experience in such situations. When we act, we respond to the original input by performing in such a way to direct our personal force as a fitting response to that input. So do we engage, time after time, striving to match our personal force to the input we receive and interpret. We are organic systems, so the flow of energy through us is both dynamic and biological, or *bioenergetic* in nature. Therefore the reality of engagement constitutes a bioenergetic flow between our inner and outer worlds, one we are intimately acquainted with, the other largely hidden from us, secluded as we are within our own minds. Everything—including our survival, including our happiness—depends on that flow as how we interpret and act upon it.

Reality, then, is not only bioenergetic but is constantly in flux from one moment to the next. It is nothing fixed or solid we can point to, as an electron is not fixed. Reality, a function of both our perception and our action, is a function of our actively engaging with our world in spite of being almost blind to it at the same time. Everything depends on how we construe the patterns of energy impinging upon our senses in relation to earlier situations in which we recognize similar patterns, along with the consequences, feelings, and meanings we associate with those patterns. These together give meaning to the current flow of energy through our neural network, inducing a recognizable sense of reality. We are on familiar ground, applying our skills in habitual yet meaningful ways, responding as we have learned through cumulative experience to respond in such situations. On our own, by trusting our experience to see us through, we do what we can one more time. Practice makes perfect. We get good at skills we frequently apply, proving once again that we have a firm grip on the reality of our situation.

7-5. Biosphere. Reality gauges our success in transferring energy to and from the biosphere we live in. Are we getting enough sunlight, air, water, food? Are we in good health? Are we sexually engaged? Do we have friends, colleagues, helpers, children, neighbors? Do we welcome strangers into our midst? Are we kind to animals? Are we actively productive for the well-being of our community? Do we contribute as much or more than we take from others? Do we strive to meet ethical standards in our engagements? Do we live *with* the Earth rather than just on it?

Reality, then, would be the life-affirming (or disaffirming) balance we achieve with the biosphere of our home planet.

7-6. Cultural reality. Science, religion, philosophy, psychology, economics, politics—these and other cultural institutions are fields of reality unto themselves as what have been called "finite provinces of meaning." As are art, music, sports, games, media of entertainment, agriculture, industry, education, transportation, and all the rest. As we engage opportunities within our culture, so do we become real to others—as quarterbacks, farmers, singers, chess players, taxi drivers, dancers, teachers, nurses, and the over eight-thousand other job titles that keep modern America running in good order.

One of the greatest inventions of all time was the position of referee, umpire, judge, linesman, arbiter—decider—who was given responsibility for officiating during engagements to assure fair play by all parties. Scientists have come to rely on peer review, statistical analysis, and citing sources for every statement in publishing their studies. Musicians take lessons, practice, rehearse, and attend master classes to bring out their best. Culture can be a self-help operation; it can also be vindictive as, for example, in the case of the U.S. invading Iraq to make somebody in the Middle East pay for the felling of the Twin Towers with workers trapped inside. The United Nations was intended to serve as a forum for settling disagreements, but also serves as a forum for nations to choose sides against their traditional or ideological enemies, while supporting their friends.

How much of reality is due to participation in a particular culture, whether religious, say, scientific, economic, or political? That is, how much of our take on reality depends on where and how we choose to direct our passions? On what we care about and work hard to support? Religion, I would say, is wholly cultural in nature, a product of traditional views, beliefs, rites, structures, and scriptures. It is a belief system and way of being in the world that we are born to, and come to accept or deny on the basis of personal experience. Science is a discipline we learn to apply for ourselves, so is both cultural and individual. It, too, is a belief system and way of being in the world, but unlike religion, it is a system that we can expand through our own efforts, leaving it larger at our death than it was at our birth. Science is participatory and expandable through practice, not primarily a matter of set belief.

7-7. Ritual reality. To a large extent, reality for each of us is tied to the rituals we perform out of training and habit. That is, to the routines we run ourselves through in order to feel familiar to ourselves. Which make reality a matter of performance of routines we have become accustomed to. This is how it is done, we keep saying to ourselves, applying our skills to familiar tasks. Cooking dinner. Washing dishes. Driving to Omaha. Watching the news on TV. Blasting off for the International Space Station. Playing chess. Making love. Cutting diamonds. Walking the dog.

Such routines reside in motor memories that remind us which muscles to move in what sequence with a particular degree of impetus. We don't have to think about what to do, we just do it—as long as we're not distracted by anything else, which would make us lose our place and ask, Now what was I doing? Could reality be a matter of not thinking at all, just accepting and performing the task at hand as we have internalized it? Of using our bodies, not our minds? Of putting ourselves on automatic pilot and plunging ahead? Is that the zone a priest is in when celebrating Mass? A teacher when following a syllabus? A politician when giving a speech? No!, No!, and No!, they would cry, but why are they so sure of themselves? We pay attention primarily when we are *unsure* of ourselves. If we know what we're doing, we don't have to think about it, we just do it in the real world. In the world we think of as real because we're so used to it we have no doubts about it. It just is. To us, totally real. Reality, then, comes down to how we learn to uses our bodies and our tools in relating to situations we put ourselves in. That is, reality is how we conduct ourselves in situations we have trained for. Think video-game-playing school shooters.

7-8. Background reality. Could reality be a matter of assuming we know what we're doing because we recognize what is called for in a given situation? Or a background of assumptions we make that tells us we are doing the right thing, or at least doing our best? Matching our habitual skills to such routine situations would become a matter of course. Reality, then, would be the path ahead that calls for the least effort. That is, the way things should be because, for us, they've always been that way as far back as we can remember. The sun hangs in a clear sky, traffic is light, we're well fed, so we turn on the radio and listen to whatever comes on. Not to worry. That's Alfred E. Neuman reality. The reality of the confident male who is sure he knows what he's doing because he doesn't care enough to ask deeper questions. Even brain surgeons, nuclear scientists, or politicians can fall into that trap. No, most of us wouldn't settle for a reality of background assumptions, even if we make a living at what we are doing and cannot imagine doing anything else.

7-9. Dream reality. This morning I dreamed that Barack Obama asked me to bring Michelle a new pair of stockings because she had trouble with the ones she had on. That task provided the structure for the entire dream which consisted of episode after episode detailing how I failed to live up to that request. My adventures took place in big rooms full of people, or city streets on steep hills. They began in a large room where I was seated next to the Obamas, moved to the streets where I flew in a light plane to a department store where I eventually got a sample card of crumpled stockings, into the streets again where I couldn't find my way back to the room, to a large hotel where I hoped to find a map of city streets but there wasn't one, back to the streets, where I woke up thwarted at not being able to complete the task I had felt honored to accept.

The dream was as real as any waking hour of my life. Real in being deeply felt, deeply earnest, deeply meaningful. I really tried to do my best. And really failed. The feel of the dream was indistinguishable from the feel of waking life. It was set in old-timey days that smacked of the turn of the previous century. Everything was well-used and run-down. The sense of urgency I felt was fully familiar. Climbing those steep streets was hard work. I was totally engaged in my task. After a while I felt worn out. Wholly situated in the dream, I was engrossed; there was no possibility of there being an alternative reality.

I wrote the whole dream out as soon as I could get out of bed. What I write here is a gross summary. The thing that struck me about the dream was the mental force or dedication with which I entered into the task I was given at the outset. I really carried on as I carry myself through life, pitching in, concentrating, doing my best—not for the Obamas' sake but my own because that's the kind of person I am. That's my routine.

Thinking about the dream, I am left wondering about the nature of mental force, drive, motivation. That which gets me going on a project and won't let me stop until it's done. It is the gravitational force that drives my dreams, engagements, actions in the world. What is that animal force and where does it come from? The message I take from the dream is that my life depends on just such a propulsive, animating force.

7-10. Life force. Mental force? Animating force? What is it that drives my sleeping and waking self? Whatever it is, it drives me to take pictures, to row my boat, to cook, to write, to make my bed, to scrape ice from my windshield, to shop, to dream, to love my partner, my children, my life on this Earth. That is the ultimate reality that keeps me going. Is it any different from the force that keeps squirrels darting in front of my car, geese flying through the sky, volcanoes erupting, fires burning, generations succeeding one another, trees growing, rocks eroding, planets whirling about the sun?

Think of the atoms we are made of, forged in the gut of an imploding star under tremendous heat and pressure, thrown into the universe to drift, to clump, to form planets from which life emerges, conscious life, life able to move, to sense, to reproduce, to write books about consciousness. It took a while, but the progression went surely stage-by-stage (as the dream I described in Reflection 7-9 progressed) because, under prevailing conditions, it was inevitable. It couldn't have happened otherwise. Yes, there were other fates as possibilities, but they never panned out. The force prevailed. The sun came into being, the solar system, planet Earth, our ancestors, and now us.

The same force that drives me, drives you. It drives priests, neuroscientists, drug addicts, computer programmers, child molesters, boat builders, mechanics, chefs, glass blowers, candidates for political office, Nobel Prize winners, the rich, poor, homeless, and self-satisfied. I know that force as common to my life and dreams, as you know it from yours. It drives us all. Makes us who we are. Drives our engagements round after round after round. We come into being. Are born. Interact for a time with our surroundings, circumstances, and situations. Die to this life. And move on to the next stage of our being in myriad ways, scattered as our atoms become when we can no longer maintain our mental and bodily routines and cohesiveness.

Somewhere, reality lies at the heart of that story, as active today as it always has been, doing its thing, driving the universe forward because it has nowhere else to go. It can't help itself. For neither good nor ill, it just keeps going, going, going, energetic until the ultimate fade to the universal heat death in which no energy is left. I can honestly say, the sun made me do it. Without solar radiation, I wouldn't be here. Without opening my eyes to see I wouldn't be here. Without taking in calories I wouldn't be here. My atoms, cohesiveness, consciousness, and dreams need the sun if they are to thrive. Matter and energy, certainly, and something else—the urge to life. Without that urge to back us up, we are too improbable to take seriously. Not that we were deliberately aimed at as an end state. We are only a passing fancy, a stage the universe must endure to fulfill its duties as a universe. We are but one episode out of quadrillions—this, then this, then this, then . . . .

My day, my dreams, are as different from yours as planet Jupiter is from popcorn. But we all participate in the same solar system, same galaxy, same group of galaxies, same galactic neighborhood, same universe. Driven by the same force to do our subjective best by the urge within us. To live, engage, and endure as long as we can. That force acting over time is our common reality. Right in our midst, the universe has an excess of energy and is looking around for something to do. It is evident in our activities at every instant of our lives. Since we temporarily possess some degree of self-awareness, we each bear witness to that force for the time we are allowed to be aware of ourselves being aware. When we wink out, others carry on for themselves until they too wink out. Beyond that I cannot imagine.

Which leaves the possibility that reality is entropy, the exhaustion of universal potential. We all age, get tired and worn down from the hassle of living a life. Maybe the life force, too, gets tired after a while, putting quits to the enterprise begun with the Big Bang. I offer that thought as a reminder to make the best of every engagement left to us while we still have a chance.

7-11. Comparative reality. In the beginning was comparison between what was and what is or could be. The disparity between them makes all the difference: right or wrong, good or evil, mad or glad, approach or retreat, big or little, hot or cold, up or down, red or blue, smooth or rough, young or old, sick or well, love or hate, win or lose. The ability to make such distinctions lies close to our hearts, that is, to our passions, the forces that drive us ahead. We prefer one choice to the other. We shun compromise and middle ground. Our senses sharpen distinctions so that similarities and differences stand out, enabling us to make decisions one way or the other. This belongs with this; that doesn't fit the pattern. Think political differences, religious differences, ethnic differences, sexual preferences, team preferences, dialects, fashions, musical genres, cuisines—wherever we can tell the difference, we tip the balance to one side or the other.

Our passions and distinctions reflect our subjective versions of the life force. For better or worse, they reveal the truth nearest and dearest to our hearts by steering our engagements this way or that. I approve of this; I don't care for that. This pleases me; that distresses me. Comparison makes these choices possible, and passion determines the balance between the two. It is no fluke that our language offers us so many choices: in or out, over or under, east or west, fast or slow, near or far, big or little, heavy or light—and pays such slight attention to gradations in between. We have trouble with ambiguity; we like things clear and bold. That's how we are made. Either-or is the way of the real man, not some wishy-washy wimp. Fish or cut bait; chief or Indian; with us or against us. Don't mess with Mr. In-between. Which, in a world where almost anything is possible, leads us to parody ourselves by seeing everything as either black or white. Moonlit seascapes on black velvet would seem to be our idiom of choice. But we do that to ourselves in order to make our choices that much easier. Subtlety is a philosopher's game, and most of us don't have the time or energy to worry such minor distinctions as are the boon and bane of philosophy.

We survive by our wits, and our wits have two basic positions: on or off. Yes or no. Go or no-go. In wedding ceremonies, the pact is sealed by two people each saying "I do," thereby legally committing themselves to each other for the foreseeable future. Contracts are signed statements of commitment. We believe they are in our own best interests. At the time of signing, the life force is evident in our pledge and our passions. At the time, they are our ultimate reality, as my urge to provide stockings to Michelle Obama was the whole thrust of the dream I recounted in Reflection 7-9 above.

Where do such urges come from? From the depths of our being who we are. All the way back to the nova that burned itself out and had no option but to collapse, in the process bringing the atoms our bodies will be made of into being. Having children stems from the same urge, from necessity. We often have no notion of any consequences that might flow from what we are doing, but we commit ourselves nonetheless. If we knew then what we know now, it might never have happened. We forge our destiny and the destiny of others in light of our most passionate urges. The life force speaks through us at such times. That is the reality to which we are born. The inescapable logic of collapsing stars and unbridled (even if attractive and decorous) passions. Something will come of this. It always does. But what the consequences might be we have no way of knowing. The urge is all. The tension, the disparity is ours to deal with.

Consciousness, as I now believe, is sparked by such comparisons as that between what *was* and what *is*, or what *is* and what *might be*. The life force uses such comparisons to assert itself one way or the other. When hungry, we look for something to eat. When thirsty, we drink. When short of breath, we inhale. When afraid for our lives or our dignity, we run or make a stand. When sexually aroused, we make love. When concerned for our children or grandchildren, we intervene. When idle or lonely, we look for projects or relationships. These are real acts with real consequences. Life doesn't get any more real than that.

7-12. Reality as repertory of patterns. Consciousness flows from a mind rooted in the past having to come to terms with a present situation. Three neural substrates are engaged: the seats of memory, perception, and the life force. What's happening, we want to know. What does it mean? What should I do? Current sensory impressions are compared to patterns familiar from the past, discrepancies noted, judgments made, and we—our situated selves—engage the current situation one way or another as we feel is appropriate to our well-being and survival.

The present moment is known to us as a field of ambient energy impinging on our senses. The meaning of the moment is based on our fitting the effects of that energy on our senses to patterns preserved in dynamic circuits in our brains. In my case, a sensory impression activates a particular circuit within my network of possible circuits, a resonance is found (or not found), making the incoming signal familiar (or novel). A comparison between incoming and resident signals results in a sense of parity (or disparity), to which I ascribe a particular meaning, leading to a course of deliberation and subsequent action.

I am no neuroscientist (as you know), but that is roughly the scenario I find playing out in my head time and again as I observe the workings of my own mind. I see myself as a walking repertory of sensory patterns available on call at any moment to review incoming patterns for their similarity to patterns on file gleaned from my life experience. Depending on similarities and differences between my repertory of patterns and an incoming signal, I determine what that comparison means in light of my earlier experience, and then face the challenge of deciding what to do (if anything) as guided by a life force that inclines me one way or another.

If a sensory pattern proves to be unfamiliar, I am thrown back on whatever resources I have for associating it with particular episodes of experience in earlier days. I can free associate, trying to locate something—anything—I am reminded of, much as I run through the alphabet in trying to recall a name to go with the familiar face walking toward me on the street. If the face doesn't register in time, I can smile and pass on none the wiser, or, hoping some clue will emerge, say, "Hi there, how's it been going since last we met?" I might even admit I'm at a loss and say, "Remind me who you are."

We all know people whose perceptual skill is legendary, or their skill at remembering, or even their resourcefulness in devising strategies for linking the two orders of consciousness (matching faces to names, say, or names to phone numbers). In my own case, I am well aware of how fallible I am when put on the spot and am asked to name a particular bird, to recall an event I lived through but have let silently slip into oblivion, or to rattle off the pros and cons in reaching a particular decision. I am not systematic in my mental habits but rely more on whatever system spontaneously comes to mind. As I do in writing this book. Rather than consult the literature in the field, I plunge ahead by putting myself in a particular situation and seeing what comes to mind. The dream I mention in Reflection 7-9 came to me on the morning I sat down to write the next section of this chapter. It made sense to write about that dream *as if* it were trying to tell me something about reality. I am always open to signs and suggestions.

My strategy is to immerse myself in my own introspective findings, and trust my unconscious mind to lead me to the next step, and the one beyond that. I generally reread what I wrote the day before, and ask myself, Where does that lead? I am writing my own book here, not rehashing what others have found before me. What interests me—and has interested me for over thirty years—is the workings of the only mind I have immediate access to in terms that are personally meaningful, rather than the systematic understanding of consciousness through animal studies, psychotherapy, philosophy, trauma, third party interviews, brain scans, pharmacology, and all the other ways we have for grappling with the workings of mind and brain.

I feel I owe an accounting of my own mind to the world as one particular case in point, rather than chase after a universal understanding of all minds in all conditions under all circumstances. I leave such studies to those who are drawn to them. My preference is to use noninvasive means under everyday conditions. The result is I find myself writing about aspects of mind that others don't seem to find interesting. The reverse is obviously true, my neglect of conventional wisdom. I let my personal life force serve as my guide in following from one insight to another, inventing a vocabulary for discussing my findings as I go.

Quakers have a saying about finding their way in the dark: "as way opens," they say, and though that way is unmapped and not at all obvious, eventually it does appear once you get accustomed to the woods (or whatever it is you want to work your way through) by a process of immediate engagement driven by the life force. That's where I find myself in writing this chapter, as I picture Dante wending his way through hell, purgatory and paradise. Not that I am any Dante Alighieri, but our episodic, step-by-step methods share some similarities. I am Dante and Virgil—self and guide—in one body.

7-13. Situational reality. Situations have a bad name among those favoring universals and absolutes, but my life, I find, is largely episodic or situational, and my actions are appropriate to situations, not abstract principles. Situations are experiential, not conceptual, cognitive, or abstract. You have to engage with them to appreciate their qualities and possibilities, as I in my boat have engaged Taunton Bay for much of my life, using my muscles and senses—not words—to build memories and make impressions on my mind so that I come to understand my situation in very personal terms.

Situations are where we locate ourselves in our minds. They are localities, but not just places for they refer to the dynamics of our engagements in those localities as well. Situations include the emotional undertones we feel, the social dynamics, the values and passions we mobilize in facing into them. They are where the life force confronts particular events, inciting us to act one way or another, depending on how we construe those events in our minds. Situational ethics are relative to that construction—to our personal engagement then and there—not to some higher, universal set of values or principles to be deployed on any and all occasions.

I locate myself in situations because I never know if I am right in how I construe them or not. Situations are sure to change; absolutes are forever. I am not a forever sort of person. I make the most of the moment and move on, learning as I go. I am a perpetual learner because I have been in a good many situations that have taught me a thing or two. My approach is to keep my eyes open and pay careful attention. I may not know anything for certain, but on a sliding scale, I understand a good deal. Which is how I am writing this book, placing myself in one situation after another, seeing what my mind can supply from that subjective placement. Something, at least, always something that opens new doors.

7-14. Alternative realities. If my particular atoms had been assembled as a least sandpiper, I would experience least sandpiper reality on tidal shores lined with tangles of rockweed, eelgrass, kelp, ideal habitat for springtails, amphipods, isopods, and other delicious tidbits. I would be situated in—and a creature of—three worlds at once: salt water, damp land, and moist air. Water would be the common denominator, so I would feel a special reverence for that place where those three worlds came together. If the temperature dropped low enough for water to freeze, I would abandon the frost zone for a more hospitable climate. When it grew warmer again, I would fly back to the estuary that made me who I am. Knowing nothing of politics, economics, religion, or war, I would be happy being myself in a supportive environment among members of my own kind.

If the star stuff I am made of had taken a different form, that of an ant, say, a hedgehog, walrus, Neanderthal, or stegosaurus, I would experience the world through a different set of senses in each case, construing a variant reality, engaging as the life force would have me, always striving to preserve the integrity of my own situated self as I ate, drank, breathed, reproduced, and went about my business.

Human reality is individual in every case, and distinct from all the other realities comprising the collective reality of our unique planet. As a species, we are headstrong in believing, as many do, that the Earth was made for us to rule. We assume Earth is as we take it to be, not worrying overmuch about how ants, eagles, or indigenous people might enjoy a different view. We dismiss them as animals or pagans—supposedly inferior beings—and let it go at that. But the evidence points to the situated reality of *Homo sapiens* being a minor sample of all possible realities, so deserving further consideration for the benefit of all earthlings.

We, the gem flashers, pizza gulpers, so-called sports-utility-vehicle drivers, automatic-weapons owners, video watchers, and haughty snubbers of other peoples and species, all have an obligation to imagine ourselves in the moccasins, hooves, or webbed feet of our fellow Earthlings in an effort to expand our horizons to account for the company we are born to and keep on our home planet. One interpretation of the life force is that it is meant to drive us to be self-centered in living for ourselves alone when, in truth we are not alone, never have been, and never will be alone in any situation we can possibly imagine. We are built to engage with any and all we meet along the way. If we find them strange, we can develop the skill and attitude to befriend them, to hear them out and come to appreciate their points of view. Instead, if we find them creatures of a foreign reality, we shun them, put them off, isolate them, demean them, subject them, imprison them, tame them, or even kill them to get them off our minds so we don't have to put up with any realities that might cause us inconvenience.

That is the human (if not humane) way of dealing with alternative realities. But denial of the existence of such realities is no solution to our predicament. In no case does might-make-right, even if that is our preferred solution — to apply more of the same force until we crush all who deviate from our chosen way of being true to but one out of many realities, none having absolute authority over any other. We are born to a congeries of realities, a heap, a jungle, a plethora. Which we often take as a threat, but is truly the greatest gift we will ever receive.

Nothing is more satisfying in life than working together among strangers of different families and tribes for the common good. Instead, we have developed the habit of battling against those who differ from us, even going to war, asserting ourselves to the death, not listening, not caring to listen. Yet we are all driven by the same force, the same ultimate reality of being citizens of the universe. In putting others down, we are cutting off extensions of ourselves — our own arms and legs — which we picture as an act of heroism, not self-mutilation.

Strange business. In my one life, I have managed to come only this far. To see the absurdity of the problems we bring upon ourselves. My solution for myself is to appreciate supporters of life — watersheds, trees, fungi, forests, streams, ponds, estuaries, wetlands of every kind where water is collected, stored, purified, and distributed. The wetness so valued by the sandpiper supports us all. We thrive on dampness, even in deserts where if we dig deep enough, we can find water. Our blue planet absorbs solar energy and brings us—every one—to life. How ironic it is that we are witlessly turning our biosphere into a hothouse, denying every sign of temperature rise. To many of us, that is an unacceptable reality, yet it is our very own doing, contrary to the interests of every person and most species on Earth.

Do we know what we are doing? Yes, we know. But are in denial because it is the nature of our memory-driven minds to look to the past, not risk heading into an unknown future. Which is logical, in a way. As survivors, doing more of the same makes a certain kind of sense. When in doubt, keep going at all costs, even if it proves fatal. We are living out a forlorn hope as if it were our highest reality. More of the same has got to be better. More chocolate, more sugar, more fat, more guns, more oil, more profit, more speed, more, more, more. Into the valley of Death at full gallop.

7-15. Introspective reality. What we need is more reflection on the state of the world and the state of ourselves. More introspection. More self-doubt. More looking within. More probing. More questioning. More self-examination. More true self-understanding. We need to cultivate our conscious mind as we would cultivate a garden. Getting to know it. Caring for it. Going slow. Looking around. Finding out who we are, where we are. Spending less time admiring our accomplishments or seeking praise, more time improving ourselves. By our standards, not theirs. Not letting others educate us according to *their* view of reality, but developing a world view based on our own experience. Teaching ourselves how to be at our best in the situations we actually (or are likely to) face. (*See* Reflection 1-6 *above*.)

Reality, after all, is our way of organizing a world we cannot know in itself. All we have to work with is our experience of such a world through our engagements, which we organize in meaningful ways to serve our interests. How carefully do we observe sensory patterns? How painstakingly do we read what they have to tell us about our situation in the world? How meaningfully do we fit them to our repertory of familiar patterns? How often do we check on the accuracy of our understanding? No, nobody teaches us to do these things for ourselves. Instead, they teach us to fit into *their* worlds, not our own. To become clones of their way of thinking and feeling, to mimic their passions, to act *as if* we were them. Which we aren't and can't be. True, we all have to earn a living for ourselves, but on whose terms? Each of us must decide how much of our personal integrity to surrender for the privilege of holding a job. If we must leave our values and passions at home, perhaps that is not the job for us.

Beware multiple choice tests. The answers we seek cannot be put down in so many words. As to right answers, there aren't any. There are only your answers and my answers. Nobody knows what happened during the American Revolution or Civil War. The men and women on the scene knew a little of it, but only a smattering, and they are long dead. We know even less about the Vietnam War or second war against Iraq. We can't even say why we fought those wars. We have to study other men's minds to get a perspective on that, and only a perspective, not the answer. In hindsight, we shouldn't have fought either one, at such a terrible cost in human misery on all sides.

That's the kind of situation we get into when we let other people make decisions for us without being able to check on what they do. Who governs the governors? In a democracy, supposedly the people. But just try to oversee the U.S. government. Any such effort puts you on the fast track to solitary confinement. Ask Bradley Manning, he knows. He is charged with making secret documents available on Wikileaks. But who classifies such documents? The government. Those who don't want the public to know what they're thinking and doing, even though it's the public's business. Especially when it's the public's business to know. Even in hindsight after we've moved beyond that particular situation.

Regarding reality, we have our work cut out for us by none other than ourselves. Introspection gives us immediate access to our inner struggles and uncertainties. Nobody thrives on deception and secret information. It is the tests we administer to ourselves that are important, not those we take in meeting someone else's requirements so we get a diploma or certificate to show we've gone through the process and rigmarole.

7-16. Reality as fond refrain. To repeat an old favorite of mine, here's some of what I wrote about reality a year-and-a-half ago (as quoted in Reflection 1-29 *above*). It rang true when I wrote it, and still does today. Having come this far, I offer it as a test to see whether or not it seems sound to new readers.

> Consciousness, then, depends on a working mind finding itself in a stimulating, energy-rich surrounding situation, both self and situation engaging in an ongoing exchange that can endure for one human lifetime. That exchange itself constitutes the reality of the two taken together as what we call consciousness. Reality resides neither in the person nor in her surroundings, but in the bioenergetic interaction between the two cooperating in tandem through the looping engagement they establish with each other. That, in essence, is the upshot of my introspective research. Reality comes down to our forming a secure relationship with our surrounding milieu—our niche in the universe—whether it is composed of significant others, work opportunities, energy sources, or situations in which we can make ourselves happen in the face of difficulty or opposition. Essentially ecological, reality is our term for the dynamic exchange of energy between inner and outer substrates that keeps us undead.
>
> And it is the disparity between those inner and outer states of energy—as revealed by our ability to compare them via our looping engagement—that gives rise to consciousness as an error signal waking us to our

current reality. It is the bite of that error signal that alerts us to a discrepancy in our understanding of our situation, rousing us to consciousness so we can investigate the source of that signal which runs counter to our expectations (*Consciousness: The Book,* pages 219-220).

7-17. Reality in brief. My reality is that I am of three minds—sense, meaning, action—in one body engaged with the body of "the other" in the form of a world I cannot know in itself but am bioenergetically compelled to engage with nonetheless by the life force I inherited from my parents as they did from theirs. I do not stand apart from that engagement as an observer looking down but myself *am* that engagement at every moment of sensible, meaningful, and active awareness.

7-18. Reality as illusion. My thesis is that all of us may be subject to the same tripartite progression of sense-meaning-action, combined into an ongoing engagement with the virtual antipodes of our situated selves in the world on the far side of our senses. The reality of that progression and that engagement cannot be told other than by cumulative experience gathered over many turns of the screw of personal engagement. Often as not, we turn out to be deceiving ourselves, as I deceived myself by cloaking the snowcapped Rocky Mountains in a thick layer of clouds (*see* Reflection 3-3 *above*), or by *not* seeing the bouquet of sunflowers between myself and my camera case when I went to fetch it (*see* Reflection 3-2 *above*). If seeing is believing, then reality may be something else altogether. Are Calvin and Hobbes real?, Captains Ahab and Nemo?, Adam and Eve? What about Big Bird?, Santa Claus?, Peter Pan?, Don Quixote?, Superman?, the Tooth Fairy?, Alice in Wonderland?, Sherlock Holmes?, Cinderella, the Little Mermaid? We truly engage with them, we treat them as real, but are they as real as, say, thunder and lightning, rice pudding, a scrape on the shins?

Think of the accessories and special effects it takes to convince us that a religious service is what it claims to be: the candles, chalices, wafers, incense, flowers, altar cloths, stained-glass windows, hymns, music, prayer books, costumes, bells, chants, scepters, creeds, intonations, offering plates, statuary, cushions, pews, prayer stools, recited phrases, liturgical calendars, fonts, basins, scriptures, times of day, processions, and all the rest. It makes for a great performance, but is it real? We engage with it, but is it *real*? Is God in attendance? Perhaps in our minds, but is he anywhere else? Is the World Series anywhere but in our minds?, the Super Bowl, the Olympics? How about King Lear?, Aida?, the Nutcracker? I would have to say that if they exist, they exist solely in human minds, and if it should happen that no one pays attention—like the party thrown with no one showing up—then they don't exist even there. The tree fallen in the forest isn't real until you stumble across it. All else is a jumble of assumption, memory, wishful thinking, hearsay, reputation, conjecture, possibility, uncertainty, and outright fabrication.

The high road leading directly from sensory impressions to rote routines and habitual actions makes it terribly easy for us to deceive ourselves in resorting to well-learned behaviors in novel situations. If comfort and familiarity are our guides, we will always rely on tried and true methods, whether appropriate or not. The low road to action via our situated selves offers a harder, more challenging route requiring creativity in rising to the occasion in new ways. At times when well-drilled habits and routines are not adequate, then the extra effort of thought and creative imagination is our only hope in deriving a course of appropriate action. The conservative approach is to rely on solutions already in place; the liberal approach is to tailor a novel solution to fit an unprecedented situation. Either way, we run the risk of deceiving ourselves, so must use our best judgment in deciding which route to take.

7-19. Heritable reality. I am calibrated by my culture in such a way that *time* is the medium of my sensory impressions (*it*-changes), *space* the medium of my actions (*self*-changes), and *space-time* the medium of my meaningful engagements (*mutual* changes) with the mysterious "other" I cannot know. Reaching out to the world of the other through bodily motions, my body actively exists in space. Opening my senses to that other, my body receptively exists in time. Interacting with that other in round-after-round of active and receptive engagement, my body exists in a three-dimensional helix of space-time, which I picture as the wrapped, ramping threads of a screw around a shaft driven by the life force.

Loops of engagement, that is, account for my simultaneously serving as *agent* and *object* at the same time, an agent in acting as I do, an object of others acting on me as they do, my engagements serially joining the two states of my being into one pulsing, bioenergetic reality in space-time. When I act in space, I see myself both acting and being acted upon in response, as I do when I perceive in time as well. Two modes of being in one body, one giving through action, one receiving through sensory impressions, the two together enabling coherent engagement with the world.

That, I propose, is my organic reality as I live it every day of my life. And as my parents lived it in their lives, my grandparents on both sides in theirs, and my sons in theirs. And as my grandchildren (should I have any) will likely live it in theirs. Five generations, all driven by the same force. Change and dynamic continuity at the same time. Echoes from the past resounding through five generations of sensory impressions, meanings, actions, and engagements in one family.

My grandmother Laura A. Gale Perrin died September 17, 1896, the day she gave birth to her son, who became my father. He survived the loss, but barely. Without a mother, his heart skipped a great many beats, which he made up for by working hard, and clenching a pipe between his teeth all his life. But he knew little about raising children, so in his own marriage, when he came to have three sons, he wasn't sure what to do with them. He turned to childrearing "authority" John B. Watson for advice (*see* Reflection 6-11 *above*), who advised treating children distantly *as if* not intimately related to them, which fit neatly with my father's early experience, so was passed on to his three sons, of whom I was the middle child. When I came to have children of my own, I wasn't sure what to do with them, either. I read to them, changed their diapers, and took them places in eastern Massachusetts like Walden Pond in Concord, Museum of Comparative Zoology on the campus of Harvard University, and wharves in Boston Harbor to watch the tugboats. But I was too engaged in trying to figure out who I was to help my children form solid identities of their own, and in two marriages grew to feel even disconnected from my wives, so left because I felt I wasn't real to myself in either domestic setting. Such feelings are toxic when it comes to helping children come into their own, as I didn't help mine ease into productive lives, leaving them to cope for themselves, with help from their respective mothers and friends. My eldest son became addicted to heroin and killed himself with a shotgun on his twenty-second birthday, for which I largely blame myself, even though I thought I could divorce my wife without leaving my children in the lurch. I spent weekends with them for many years, showing them the sights around Boston, but that ended when they moved to California, and then to Italy, so I simply was not part of their teenage lives. In my second marriage, having one child and one step-child, I turned into an absentee parent, still committed to finding out who I was into my mid-fifties.

Having had undiagnosed celiac disease since infancy, I had less and less energy every year. Once, when I was lying listlessly on the bed, my second wife came into the room and after five minutes said, surprised, "Oh, I didn't know you were here." I immediately saw that I didn't exist in that marriage, either. So came to Maine at age fifty-six, and set about reinventing myself yet again. I had a series of terrible rashes, which no doctor managed to diagnose. One dermatologist did a biopsy, which led to a diagnosis of *Dermatitis herpetiformis*. What causes that?, I asked. Irritation of cells in your skin, I was told, as he prescribed yet more ointments to apply to my flaming hide. Six years later, when I was sixty-five, I typed that Latin binomial into Web-

Crawler, and the first hit from St. John's University informed me my rash was caused by celiac disease. Quitting wheat on the spot, I watched the rash recede, then disappear over two-and-a-half weeks, resetting my clock. I made it to Y2K after all. In a very real sense, I am now sixteen years old.

So here I sit writing a second book about knowing my own mind, without help from anyone experienced in the field of mental life, independent as always, as my father lived out his life, my brothers are living theirs, and now my younger sons (in their forties) theirs. Reality winds down the generations as the screw turns, cycle after cycle, through one line of descent, twisting the intimate spatio-temporal engagements of all concerned into a legacy of balked engagements, stress, and deep yearnings for love and approval. How we are to engage (or not engage) is the heart of the matter. That is the reality that gets passed down. How to read sensory patterns, how to fit those patterns into meaningful situations, how to make an appropriate response to those patterns and meanings. All driven by an insistent life force as the voice of our star-planet system, true parent of us all.

In the vast spread of the universe, this is the one life I get, the one flow of consciousness. Whether I make the most or the least of it, I owe an accounting of how I have spent the days of my life. My story ends here. The rest is for others to tell. ○

Chapter Eight

TOWARD A THEORY OF MIND

8-1. Subjectivity. By definition, consciousness is subjective; it cannot be fit into a framework that insists on objectivity. The locus of the unconscious may be the brain, but the locus of consciousness is the mind, enabled by the brain, but not identical to it in part or in whole, as an electrical circuit is not identical to the copper wire it is made of. Such circuits acquire characteristics by being *turned on,* as consciousness must be turned on or *aroused.* Such effects as resistance, inductance and capacitance arise from the existential flow of electrons within circuits, specifically, from interactions within that flow itself that affect how electrical energy is received, stored, and distributed. They arise from emergent and kinetic (not static) properties of electrons moving through closed circuits under particular conditions. Consciousness is somewhat similar in being enabled but not predetermined by the brain. Instead, as I see it, it rises above neural circuitry to interact through the comparative resonance and dissonance between its several parts.

Quantum physicists incorporate qualities of mind into any observations they are likely to make. That is a huge step in the right direction. Insisting that subjective observers remain essentially pure and aloof from their personal observations is an exercise in ideology. Each observer is a multidimensional set of mental variables engaging the world in a variety of ways simultaneously. Results depend on what he or she had for lunch, whether he or she is well-rested, when he or she last had sex, and so on. When two or more scientists gather together, it only gets worse, that is, more complicated and less objective, because of the chemistry within and between them. I think a new branch of science allowing self-reflection as a productive and honorable profession *based on first-person experience* is due to emerge. This will compensate for methodological and ideological rigidity in the practice of neuroscience, allowing a more complete accounting for what consciousness may be—and how it arises from the brain—to appear at last.

8-2. A tripartite model. In everyday practice, consciousness addresses three tacit questions: 1) What's happening?; 2) What does that mean to me in my present situation?; and, 3) What should I do in response? *Perception* fields the first question, the *situated self* takes the second, and *action* resolves the third. At the risk of oversimplifying, I visualize the mind as being divided into interconnected departments or modules corresponding to this tripartite model. The perceptual department of mind extends between sensory receptors and the hippocampus, which facilitates the formation and recall of memories. Integration of sensory qualities does not pose a problem because each quality is wired into (situated within) the spatial and temporal geometry of a unique conscious self. What I call the *situated self* is at the heart of consciousness, with access to sensory impressions, understanding, memory, comparison, dreams, values, feelings, and imagination. And both of these departments—perception and the situated self—connect to motor areas of mind and brain. The situated self connects via the planning areas of the brain, the province of judgment, decisions, goals, projects, and relationships. The sensory department, too, can fire directly (and unconsciously) to the motor area, where impulse and habit can direct personal effort and force toward the outside world.

But the story doesn't end there, for by being caught up in a program of action, perception is set to gauge what happens next in order to follow through on its commitment to effective and appropriate action, revising or even countering its initial assessment. Few actions are ends in themselves; most are stages in an ongoing progression of continuous activity. As in tennis, the game isn't over once you serve the ball; you immediately position yourself to hit it again as it whizzes back over the net, and then again, and again. If you want to eat, you provision your pantry, decide what to have, prepare it, cook it, serve it, eat it, and wash up afterwards—and repeat the performance a few hours later.

I visualize personal consciousness as a process of ongoing activity that modifies our felt situation as we go, morphing time and again into a wholly new situation, which we fail to address at our peril. Survival is somewhat like tennis: we've got to keep our eye on the ball at all times. A rhinoceros could rumble out of the bushes any moment or, more likely, a child chase a ball into the road ahead. The prize goes to the ever vigilant, not merely the fast, strong, smart, and beautiful.

8-3. Loops of engagement II. The succession from perception to situation and on to action never ceases. I picture consciousness in terms of durable *looping engagements* by which any given action immediately initiates a subsequent round of perception-meaning-action until the situation itself is no longer relevant, stopping the clock, inviting other situations to take over and start a new round, spiral, or helix of engagement. This helical (because never exactly coinciding with its beginning) series is far more than a succession of working memories or hand-eye coordinations; this is how we make ourselves happen in the process of continually reinventing ourselves and our worlds.

8-4. Organ systems II. Humans did not create consciousness all by themselves; they inherited it from their distinguished ancestors who, even on the cellular level, discovered that the membrane setting an organism off from its immediate environment had to be permeable in both directions, in *and* out. Exchange (interaction, give-and-take) is the rule, not the brilliant exception. At every scale, metabolisms need to be fed from the outside, and the buildup of waste products simultaneously eliminated. Voilà: *the loop of engagement.* The same basic principle applies to our pulmonary, cardiovascular, digestive, reproductive, immune, integumentary, and nervous systems. Looping engagements do not exist apart from the organic world; they are the heart of that world. So it should be no surprise that they are at the heart of consciousness as well. (*See* Reflection 1-20 *above.*)

8-5. Polarity. Consciousness is bipolar in nature, having both an interior and exterior pole. The situated self is the inner pole, the virtual or conjectural world being the outer. When we are born, we have no idea what we are getting into. We consist of a naïve inner pole that has only its discomforts and satisfactions to go on as driven by the life force, but other than by crying or sucking, we have yet to learn how to engage in order to get more of what we want, and less of what we don't want. Mother holds us in her arms, sharing her bodily warmth, her milk, her love, whispering softly, "Don't cry little baby, stick with me and all will be revealed." We do, and it is. Since conception, she has become the primal "other," the outer pole of our existence, the first world we engage with. Our lives are the histories of the engagements that follow.

8-6. Trial and error. Every new life is an experiment to see what is effective and what not within the particular niche it occupies by means of its perceptions and actions. No one else shares those exact perspectival coordinates; we are in this life to discover how far we can travel via this singular point of being. On our deathbeds we realize our journey is done; the next leg is up to those who survive us via their own points of being. The experiment never comes to an end; it is what we share with all others of our kind to see if we can't figure out what will work to keep us going, and what won't. We have only our passionate beliefs to go by. There are no universal directions, guidebooks, gurus, recipes, magic potions to help us. We are condemned to *a life of learning by doing and believing,* hoping our subjective awareness will prove sufficient to the task. Through our parents, the universe hands us our genetic makeup and says, "See what your mindful body can do with this." The rest is up to us.

8-7. Memory II. Memory is the backbone of consciousness. Strong emotion and frequent repetition build stable connections within neural networks shaped by personal experience. Connections that aren't used don't persist. Memory gives us hope, dread, expectancy, recognition, sameness, familiarity, and a sense of the future, among other aspects of awareness. Memory allows us to look for more of the same, as well as for what is new, novel, different, and mind-expanding.

Consulting my own experience, I recognize four primary types of memory: 1) *motor* memory of routines I frequently perform; 2) *Spontaneous* (or working) memory that is fleeting, typically lasting only a few seconds; 3) *autobiographical* memory that can preserve personal episodes for a lifetime as a result of long-term potentiation; and 4) *conceptual* (or semantic) memory abstracted from the flow of experience to represent persisting types or categories of sensory patterns as based on repeated presentations within a limited range of similarity, facilitating the convenient labeling of specific impressions as *concepts* approximating one familiar pattern or another.

8-8. Inputs to consciousness. Three very different inputs support consciousness: 1) *materials* delivered by bloodflow to fuel the metabolism of body and brain; 2) *energy* imparted to sensory organs that kindle impressions to be interpreted in light of prior experience as one's proprietary awareness; and 3) the *life force* we inherit with our particular genome, the urge to breathe, eat, drink, laugh or cry, heal, rest, love, nurture, engage, and keep going against all odds. Ambient energy and adequate nutrition are basic substrates of consciousness; reducing availability of either one results in mental impairment and degradation (as in solitary confinement). Consciousness itself flows from the life force, the need to engage, to know what's happening, to make meaning, to plan and then act, and then to discover what happens next. We call this yearning to engage "soul" or "spirit," but it doesn't belong to us as individuals. Rather, it is the endowment we receive by being born as organic beings to an energy-rich planet that offers us a toehold in the universe.

8-9. Levels of consciousness III. Within the brain, two basic routes are available for passage from sensory impressions to appropriate actions: the first is a direct and unconscious route of reflex-mimicry-habit-routine-custom-belief that prompts immediate action on appearance of particular sensory cues; the second is a longer and slower route of conscious consideration that entails reflection, judgment, and decision in arriving at a plan of action situated in subjective life experience. Both impulsivity and consideration are available to us in every situation. We choose between them on the basis of our self-awareness as actors in a world largely of our own making. If we size-up our situation incorrectly, that is our call and our error. If we want to be sure of doing the right thing, we must examine the situation carefully to increase the probability that what we do is appropriate to the specific set of circumstances we are in. I refer to these two op-tions as being on different mental levels, the unconscious and the conscious, what I have elsewhere referred to as the high road and low road (*see* Reflection 3-19 *above*).

8-10. Animal consciousness II. Other kinds of consciousness become apparent from observation of animal behavior. In many species, individuals are apt to be differentially affected by sensory stimulation (depending on genetic, dietary, experiential, physical, developmental, and social variables, among others), and to exhibit idiosyncratic behaviors as a result. Speaking more generally, different species live in different sensory worlds, and appear to be conscious in a variety of ways. Humans lack the lateral-line receptors of fish that detect the relative motion of water against the two sides of their bodies, allowing them to orient themselves in a current, and to detect unmoving objects at a distance. We don't have the hearing sensitivity of bats, scenting ability of dogs, sensitivity to heat of pit vipers, directional hear-ing of deer, scanning ability of electric fishes, magnetic sensibil-ity of eels, sharks, and birds. We may be fellow creatures, but our respective sensibilities situate us in very different niches in parallel worlds of consciousness on the planet we share. (*See* Reflection 1-32 *above*.)

8-11. Comparison II. *Change, difference, motion,* and *comparison* are other basic principles underlying consciousness. Memory not only allows us to categorize sensory patterns, but also to notice what has changed or is different in respect to their former make-up or to a set standard pattern. Comparison of neural signals in, say, adjacent or reciprocating cortical columns creates a sense of relationship (depth perception, sound location, symmetry, consonance, dissonance, extension, expansion, proportion, opposition, elaboration, and so on) in consciousness. I view *comparison* between current and prior impressions as firing up consciousness itself in proportion to the disparity detected. If nothing has changed, there's no need to pay attention and we can get by on habit and routine. But if changes are noted, are they for better or worse? We spend much of our mental energy evaluating implications of situations that change and develop.

This suggests to me that consciousness is a form of memory, or, more accurately, *a way of remembering* in a current situation so that conventional and novel impressions are compared, and any disparity directs attention to discover what if anything can be told by the difference, as in similes, metaphors, and analogies. In other words, discrepancy is viewed within a framework of sub-jective meaning, enabling evaluation of what difference it makes. When we hit a wrong note or use an unexpected word, it is the difference from what we expect that gives us something to work with in either correcting ourselves or being creative. The motto of every educator might well be: *Vive la différence!* (*See* Reflections 1-23, 5-15, *above.*)

8-12. Meaning II. Each individual stream of consciousness is *unique* and available to only one specific animal or person. In that sense, each conscious being has a proprietary interest in its ongoing experience within its experiential niche, and is person-ally responsible for actions based on that experience. *Meaning* is another fundamental principle of consciousness, evaluating the new in reference to the expected or commonplace. Each of us survives on the strength of how well we interpret the flow of energy through our sensory portals in light of our prior experi-ence. The meaning of a sensory pattern is not conveyed by the pattern itself but by how we subjectively construe it. It is invented on the spot, not given by others. Meaning is a product of assimilating sensory impressions to the existing order of sub-jective understanding, or if that doesn't work, of expanding that order in such a way to accommodate novel impressions. (*See* Reflection 5-21 *above*.)

8-13. Time and space II. Comparisons resulting from our ways of believing and remembering lead to detection of discrepancies, which are changes since we last looked (listened, touched, tast-ed, sniffed). Perceptual changes noted by a passive observer (as when sitting still listening to music) are changes in *time*; by a moving observer (looking out the window while riding along in a car or bus) are changes in *space*; by an active and moving ob-server (dancing, climbing a tree, bushwhacking through woods), changes in *space-time*. Time and space aren't out there coursing through the universe, they are in us as a sense of calibrated change. Our culture provides the calibration; we provide the awareness of detecting and enacting change. When the cultural calibrators die off, only change will remain, and when individual memory dies, awareness of change itself will fade away. (*See* Reflection 1-18 *above*.)

8-14. Phenomena. The aim and purpose of consciousness is to achieve *behaviors appropriate to one's actual situation in a world that cannot be known in itself*—a logically impossible task, but one we attempt at every waking moment. Mind is an emergent property of the brain, but the workings of the brain in terms of the electro-chemical traffic flow through idiosyncratic neural networks are very different from the workings of the world outside our bod-ies, so sensory impressions are not simply representations of the world but point-for-point creative renditions in what amounts to a singular universe within consciousness. In practice if not in convincement, we all are dedicated phenomenologist because phenomena (appearances, impressions) as rendered by our sen-sory systems are what we have to go on, not things in them-selves. Since each being is unique, its stream of consciousness is unique, and the world it construes for itself is unique—its actual situation being a matter of conjecture and imagination based on the evidence of its senses in light of its situated understanding.

8-15. Dreams. *Dreams* and *reveries* are variations of consciousness in which we are shut off from the world of conventional action and stimulation, but can nonetheless *simulate* sensory impressions courtesy of random eye movements and fixations that activate neural pathways to stir up fleeting images from memory *as if* we were fully awake. Our dreamselves cannot engage, for they can neither perceive nor act, so we must make do with memory, letting our dreams themselves illuminate the journey of the self we are, without being situated other than in our personal histories. As *potential* perceiver and *potential* actor, the dreamself is at the core of the waking self. We do well to pay close attention to our dreams as informants about the history of our core selves all the way back to infancy when our deeds and impressions lie ahead of us. This latent, so-called theory of consciousness is the narrative told to me in my dream-like reflections, and I am sharing with you as a gesture of neighborliness.

Here I invite you to discover the roles you may play in my dreams, much as Barack Obama and other world leaders invite us into their dreams, as the Koch brothers and Karl Rove invite us into theirs, along with writers, filmmakers, singers, musicians, dancers, athletes, politicians, rich and poor, and the rest of humanity. Who but a dreamer would concoct such a tale as U.S. military interventions in Southeast Asia and the Middle East, as clashes between tyrants and citizens in the so-called Arab spring, as the absurd filibuster rules in the U.S. Senate, as profit-driven scams by all parties in the recent financial meltdown among Wall Street's extended family?

Or for that matter, who but a dreamer would tell tales such as the Exodus and other narratives woven into Hebrew scripture at the time of the Assyrian invasion and occupation in the late eighth and early seventh century BCE? "[T]he biblical narrative is so thoroughly filled with inconsistencies and anachronisms,

and so obviously influenced by the theology of the seventh century BCE writers, that it must be considered more of a historical
novel than an accurate historical chronicle (page 175)." "[I]n
what can only be called an extraordinary outpouring of retro-
spective theology, the new, centralized kingdom of Judah and
the Jerusalem-centered worship of YHWH was read back into
Israelite history as the way things should always have been"
(Israel Finkelstein and Neil Asher Silberman, *The Bible Unearthed*.
New York: The Free Press, 2001, page 249). Not only read back,
but put forward as scriptural truth for all time to come, yet
another self-promoting, other-degrading, system of tribal belief.
Or like Christianity and Islam, another collection of stories
meant to appeal to the vanity of those who see themselves as a
breed apart, a higher form of humanity.

Even more to the point, who but a dreamer would come up
with the model of consciousness I am striving to depict in this
book? Sanity sleeps as our dreams rule the world, not metaphorically but through our actions. As Goya put it, *"El sueño de la
razón produce monstruos,"* The dream of reason begets monsters.
Or as Italian playwright Luigi Pirandello wrote, "It is true if you
think so." It's not so much that dreams are fantastic as that *so-
called reality is largely a dream enacted and become ritual.* Think
fracking, school shootings, global warming, human trafficking,
religious strife, sexual abuse, torture, child soldiering—whose
dreams are these? Not mine or yours, perhaps, but they are all
somebody's dream. The lies we live are much like dreams.

8-16. Introspection II. Science, I think, traditionally underplays
the virtue of *introspection* as a message from the interior of one
person. The art of introspection is in accepting whatever appears, not judging or dismissing it beforehand because it does
not meet designated research criteria. The arts and humanities,
on the other hand, along with many other creative aspects of
human culture, *celebrate* individual differences, and play them
up as valuable in themselves for distinguishing us one from
another in admirable ways. If we were all the same, we would be

replicas, and life would progress from dull to duller to dullest. (*See* Reflection 1-6 *above.*)

8-17. A tale of two selves. The upshot of this narrative is that we are heavily invested in our subjective consciousness as the lived edition of personal survival, that tale of two complementary centers—subjective and virtual—facing off against each other at opposite poles of our engagements, separated by the membrane that serves as our skin. This is a tale of two *selves,* for the virtual world we imagine is largely fleshed out by our own experience as we remember it, so is an extension of our situated perspective as a kind of alter ego accompanying and complementing us in our experiment to see if we can't at least get some things right. Which we all manage to do as demonstrated by our ever spiraling engagement in the streaming adventure of mental life, giving others the impression we are present and accounted-for. To those others, we serve as the virtual poles complementing *their* subjective selves as situated in the shadows of their own impressions, meanings, dreams, life force, and actions.

8-18. Above all. Wherever on Earth we are born—East or West, North or South—we are born *of* woman, and born *to* the stars. We are Earth made flesh looking upon a vista of lights in the sky. From the outset we are of two natures, mortal and celestial. Our lower nature is hard to bear because so fraught with peril, temptation, and suffering. Our higher nature is stunning in its harmonious radiance and diversity. Our task in each case is to balance our two natures, not to revel exclusively in either one or the other. This creates a tension within us, a yearning for both familial warmth and heavenly illumination combined in one life.

Which is what both science and religion strive to inspire: a unity of nurture and illumination in passionate embrace—religion by building on the past, science by opening to the future. Is that too much to ask? Love and awe together in one vessel? But that is precisely what the life force drives us to seek at each instant. As Buddha sought to master sickness, old age, and death through personal discipline. As Moses sought to bind his people under one law. As Jesus sought to discover loving harmony in the face of human diversity. As Mohammed sought to unite the fervor of three hundred sects into one ardent system of belief. As scientists are struggling today to reconcile four fundamental forces within one flash of coherent illumination, a task that yet evades them. As we all strive to achieve in our individual rounds of engagement, balancing our situated selves with our virtual worlds at each moment, being wholly ourselves in a universe wholly other, promoting individual happiness and the general welfare at the same time.

Even as we know it is impossible to attain such a balance in any truly enduring sense, we also know we have no choice but to keep trying out of deference to our parents (genetic heritage) and the cosmic order that led to our birth.

8-19. After all. Yes, we are unique individuals, and at the same time, members of families, tribes, peoples, and a particular universe. I have emphasized the subjective nature of my mind because that, I feel, is the level of consciousness I am (and I imagine we all are) apt to pay least attention to, so least understand. In the end I must acknowledge that in addition to our being ourselves, we all belong to higher natural and cultural systems of organization. In each case, our *belonging* is every bit as important as *being* who we are as individuals. My self is composed of particular organs and organ systems. When my liver gets sick, I get sick. When my brain suffers a stroke, I get confused. When my organ systems fail, I die.

Beyond my embodied self, I belong to an extended family that includes my parents, siblings, several spouses and partners, my children, and other relatives. Beyond my family, I belong to a tribe made up of bands of more-or-less similar families which, taken together, constitute a people. And every people belongs to its planet, solar system, and universe. Between my selfhood and layers of membership, I find I am a congress of organic beings, most of whom I would not recognize if I met them face-to-face. From the point of view of a bacterium on my skin or in my gut, I am a universe it takes wholly for granted, even though at every second its wellbeing depends on my mortal jelly.

When I say my individual self is situated between my perceptual and behavioral capabilities, that is true as far as it goes. But I am also embedded within a family, community, species, and universe, each level far more complicated than I can imagine, much less contemplate. Somewhere there is an asteroid of some size orbiting the sun, an asteroid fated to collide with my exact coordinates on the surface of the Earth. There are similar future impacts heading my way in the context of my family, tribe, people, and geophysical coordinates. My being and belonging are functions of a universe containing, in addition to my mind, a host of such dramatic features as shooting stars, auroral displays, blossoms, nursing babies, plate tectonics, epidemics, collisions, extinctions, and all the other greater or lesser wonders and catastrophes we are subject to.

In the end, there is far more to our being and belonging than we can possibly grasp. I offer these reflections as so many petals blowing on the wind. While they last, I can only enjoy their swirling dance. Once free of my mind, their fate is their own.

Now that our journey together is over, I am trying to introduce myself before we part to give you a snapshot of your companion through these several chapters. I am he whose autobiography unfolds within a certain set of drives and feelings between overlapping motor and perceptual capabilities within a distinct body embedded in a unique family and its culture on a planet whirling around a nearby star in a galactic neighborhood and its universe.

That's me, the one who does his or her best to engage the layers collectively making up such a situation while I briefly have the chance. And as near as I can tell, that's also you.

Voyage of a lifetime, it has been a great trip. I can only wish you well from here on. O

INDEX

O

www.ingramcontent.com/pod-product-compliance
Lightning Source LLC
Chambersburg PA
CBHW022017090426
42739CB00006BA/171